GLOBAL AGRICULTURE DEVELOPMENTS

MONOCULTURE FARMING

GLOBAL PRACTICES, ECOLOGICAL IMPACT AND BENEFITS/DRAWBACKS

Global Agriculture Developments

Additional books in this series can be found on Nova's website
under the Series tab.

Additional e-books in this series can be found on Nova's website
under the eBooks tab.

GLOBAL AGRICULTURE DEVELOPMENTS

MONOCULTURE FARMING

GLOBAL PRACTICES, ECOLOGICAL IMPACT AND BENEFITS/DRAWBACKS

TAPAN KUMAR NATH

AND

PATRICK O'REILLY

EDITORS

nova publishers

New York

Library of Congress Cataloging-in-Publication Data

Names: Nath, Tapan Kumar, editor.
Title: Monoculture farming : global practices, ecological impact and benefits/drawbacks / editors: Tapan Kumar Nath and Patrick O'Reilly.
Other titles: Global agriculture developments.
Description: Hauppauge, New York : Nova Science Publishers, [2016] | Series: Global agriculture developments | Includes index.
Identifiers: LCCN 2016017061 (print) | LCCN 2016021367 (ebook) | ISBN 9781634851664 (hardcover) | ISBN 9781634852128 ()
Subjects: LCSH: Cropping systems.
Classification: LCC S602.5 .M66 2016 (print) | LCC S602.5 (ebook) | DDC 631.5/8--dc23
LC record available at https://lccn.loc.gov/2016017061

Published by Nova Science Publishers, Inc. † New York

CONTENTS

PREFACE

With the increasing demand of food for rapidly growing population, the agricultural land uses had undergone remarkable changes round the world. In order to boost up the agricultural yield monoculture farming has been proliferated across all regions of the globe. This type of farming systems has converted most of the diversified agricultural ecosystems into high-yielding single cropping system. Even though land productivity has been increased, but this is at the cost of local biodiversity loss. The Earth has lost its rich biodiversity and human community is lacking many ecosystem services. These have huge impacts of global climate too. This book describes some of important monoculture farming systems in tropical countries that has potential socio-economic and environmental impacts in the regions.

Chapter 1 describes policies and institutional interventions for agricultural development of the backward Northeastern region of India. Various national and state agricultural development agencies, especially the commodity boards have been constantly engaged in the development of commercial agriculture in the region. Among the commodity boards, the interventions by the Rubber Board have been quite significant in terms of social and economic impacts and the entire NER is emerging as the 'Hub of rubber production' in the country accounting for 18 percent of the total area and 6 percent of the total production during 2013. The process of rubber expansion has been quite dramatic, even surpassing the growth observed in the traditional rubber growing states in south India, including Kerala, Tamilnadu and Karnataka. The phenomenal growth of rubber plantation areas in the NE region is also a matter of serious concern, as rubber is highly promoted as a monoculture system in the otherwise diverse agro-ecological environments of the NER. Rubber expansion in the NER has been justified on the grounds of: (a) ever increasing domestic demand for natural rubber from the manufacturing sector (dominated by tyre industry); and (b) the saturation of agro-climatically suitable lands in the traditional regions, especially, Kerala. Moreover, promotion of rubber cultivation in the NER has been considered to have greater social impacts in terms of rehabilitating the erstwhile shifting cultivators in the region and thereby leading to their social and economic empowerment. However, it emerges that the specific policy and institutional interventions have been oriented towards promoting rubber as a monoculture system in the NE region, which seem to pose potential threats to the co-existing land use and resource management as well as integrated livelihood systems followed by the rural households, dominated by tribal communities in the region.

In this backdrop, the Chapter makes a critical analysis of the compatibility of the policies and institutional interventions historically followed by the Indian Rubber Board in the wide-scale promotion of rubber monoculture in the North Eastern region in India in a comparative perspective of the traditional rubber growing regions in South India, with particular reference to Kerala. If the authors examine the trajectory of development of rubber plantations in Kerala under the institutional interventions spearheaded by the Rubber Board, it emerges that the Board had promoted a system of rubber production that was highly oriented towards monoculture without considering the crop promotion from an integrated farming system perspective. The paper further argues that given the agro-ecological diversity and the specific socioeconomic, ethnic and institutional settings as well as the pattern of livelihoods followed, the policies and institutional interventions for rubber development in the NER invariably requires an integrated and holistic approach, so as to minimise the damages caused to the fragile agro-ecosystems of the region and thereby not disturbing the inland fishery, livestock and forestry integrated livelihood systems in the region. Replication of the rubber monoculture as had been widely promoted in the traditional regions, especially, Kerala to the NER, can be a cause of conflict with the pre-existing as well as coexisting agricultural production (including food crops) practices/ farm integrated livelihood systems. Moreover, the institutional makeover, including infrastructure support of the Rubber Board in the region also require major restructuring to evolve an integrated approach towards rubber development along with promotion of other livelihood and rubber integrated agro-forestry systems.

Chapter 2 - The demand for sawn timber or hardwood is growing rapidly in Sri Lanka with the growth in population, economic development, and external trade of timber industry. Teak has become one of the most valuable hardwoods in Sri Lanka, which is paid premium prices at local and international markets due to its beauty, strength, durability and weather-resistant qualities. Teak growth is rapid and there is a relatively rapid turnover and produce homogeneous economic crops. The benefits of teak are in the form of goods: timber, poles, etc., and services: amelioration of microclimate, watershed, reduction of soil erosion, provision for shelter and shade, etc. Commonly, teak monoculture plantations in Sri Lanka has often been criticized for focusing too much on the management of timber and for insufficient attention on ecological functions. Generally, monoculture forest plantations damage environment including the soil, water, and even the climate on an unprecedented scale. However, the commercial teak timber monoculture plantations have a number of positive as well as negative local environmental, social, and economic impacts. As the demand for teak continues to increase in Sri Lanka it is practically vital to investigate the teak monoculture plantations in terms of socio-economic possibility as well as a sustainable solution for growing environmental issues. Hence, the study attempts to review available literature and analysis teak monoculture plantations in Sri Lanka in terms of: teak monoculture plantation strategies; merits and demerits of monoculture teak plantations: and sustainable solution for growing environmental issues of monoculture teak plantations. The main teak monoculture plantation strategies in Sri Lanka are identified as modern industrial plantations; farmer woodlots programs; and home-garden development programs. The merits of monoculture teak plantations are discussed in terms of: reclamation of degraded lands; select most profitable tree specie; greater timber yields under less pressure from other species; uniform tree structure; assure adequate supplies of timber; employment and income generation; foreign exchange savings; conservation of natural forest; carbon sequestration and capacity-building for sharing GHG benefits. The demerits of monoculture teak plantations are analyzed through: fail in protecting biodiversity in forest ecosystems; limitation to produce food with shrinking terrestrial resources; vulnerability of monocultures to disease and insects, and greater instability of production; face risks of fire; little or no forest conservation; greater risk under changing environmental conditions; impact on the environment on an unprecedented scale; negative impact on local communities and their livelihoods; and prediction of excessive supply and price declines. Sustainable solution for growing environmental issues of monoculture teak plantations are suggested as: agro-forestry initiatives; introduction of silvopastoral systems; popularization of home-gardening; and managed biodiversity protection functions.

Chapter 3 - The first attempt of raising forest plantations in the Chittagong Hill Tracts, Bangladesh was made in 1871 through importing seeds of teak (*Tectona grandis*) from Myanmar. Since then, plantation forestry has become a part of the overall clear felling followed by artificial regeneration with commercially valuable species in the hill forests management systems of the country. The natural multistoried forests of Bangladesh were under the process of conversion from natural to artificial through the removal of existing growing stock and planting with more valuable species and also with fast-growing and firewood species. Today of the total forest area, 84% (12, 04,000 ha) has been classified as natural forest and 16% (237,000 ha) as plantation forests. The plantation forests are composed of long rotation species (131,000 ha), short to medium rotation species (54,000 ha), mangroves (45,000 ha) and the rubber plantations (8,000 ha). To mitigate the demand of timber and fuel wood in Bangladesh, a priority program of introducing fast growing tree species was taken up. Extensive trials of *Pinus*, *Eucalyptus* and *Acacia* species were done to find out the very fast growing exotics for some difficult sites. After successive elimination, provenance, growth and yield trials, some promising species, e.g., *Eucalyptus camaldulensis* and *Acacia auriculiformis* were recommended for large scale afforestation/ reforestation programs especially with an intention of rapid greening the barren hills and marginal lands. Social forestry concept was getting priority with the involvement of local people and appears a successful program in many areas. However, the productivity of the present planting stock in Bangladesh is extremely low in comparison to regional and international standards. There are also reports that 20-30% of all plantations established during the last 40 years have been destroyed, and of those surviving, the stocking is much less than the expected standards. The plantations appear likely to produce well below the capacity of the site due to lack of quality planting materials and proper maintenance. All this has contributed to very low growth of 2.5 $m^3\ ha^{-1}\ yr^{-1}$ in areas where 7.5 $m^3\ ha^{-1}\ yr^{-1}$ was the original standard. The paper examines the pros and cons of plantation forestry in Bangladesh and the contribution of plantation forestry in conservation of the remaining natural forests of Bangladesh.

Chapter 4 - Palm oil is a ubiquitous ingredient found in a wide range of consumer products, including cosmetics, food and beverages, and as a substitute for fossil fuels. The rapid expansion of oil palm in the last few decades has resulted in widespread deforestation, biodiversity loss, and carbon emissions in the tropics. It has also negatively affected the livelihoods of many local communities. However, oil palm production is a highly lucrative business and the oil palm industry has brought about positive impacts on rural development, regional economies as well as lifting small farmers out of poverty. This chapter pulls together the literature on the broad environmental impacts of oil palm expansion and present an overview of the most recent research which are related to the loss of natural habitats, biodiversity, and ecosystem services. Then take into account of geographical and managements contexts, and analyse how current sustainability initiatives are addressing these issues. The analysis shows that the most developed oil palm sustainability initiative, a voluntary certification scheme known as RSPO, addresses to different degrees the majority of the impacts identified from the expansion of this crop at the plantation level. However, because of the broad range of requirements, the uptake of this certification scheme among smaller producers has been low. Furthermore, certification schemes have a limited role to tackle the broader impacts of oil-palm agriculture at larger scales. It is argued that additional mechanisms that promote an integrated land-use planning approach to oil palm landscapes and producing regions are needed, if the negative cumulative impacts from these initiatives are to be minimized and the positive benefits maximized.

Chapter 5 - The objectives of this chapter are 1) describing swiddeners' perceptions on land-use decision-making and land-use strategies in Besiq Village, West Kutai District, where development of large oil palm plantations are planned, and 2) determining the state of oil palm cultivation in Paser District, where oil palm plantations were developed more than 20 years ago, then examining the problems of oil palm cultivation in the eyes of swiddeners.

The people in Besiq Village evaluated land uses such as rattan, rubber, and oil palm cultivation based on three criteria: "return on labor," "flexibility of land and resource use," and "autonomy." They selected rubber cultivation as a source of regular cash income because of the high "return on labor," and also maintained rattan cultivation as a source of cash income for emergencies, and as material for baskets because of its high "flexibility of land and resource use." Villagers hesitated to introduce oil palm plantation development because of low "flexibility of land and resource use" and "autonomy," although they had expectations for a high "return on labor." Therefore villagers agreed to plantation development in the customary forest of the upper river area, but refused it in the vicinity of the village, where land is de facto owned by villagers. This suggests that the villagers adopted a "mosaic livelihood strategy," whereby they try to combine rattan cultivation and rubber cultivation in a complementary fashion, and accept the development of oil palm plantations partly in an effort to avoid social concerns. Based on the realities of oil palm cultivation in Paser District, the problems of oil palm plantations were examined in terms of the three criteria.

Chapter 6- Oil palm plantations are rapidly expanding along with negative environmental and social effects, such as irreversible large-scale deforestation, biodiversity loss and wide-ranging land disputes in Indonesia. This chapter addresses some of the aforementioned problems by considering alternative development strategies for large-scale oil palm plantations, focusing on the smallholder development policies and institutions in Indonesia. The Nucleus Estate and Smallholders (NES) and Project Management Unit (PMU) schemes have existed since the 1970s during the era of Suharto's authoritarian regime. The former was supported by estate companies and mainly applied to oil palm plantations, while the latter was supported by government project management units and mainly applied to rubber and coconut plantations. The important aspect here is that the PMU scheme can avoid rapid and large-scale land expropriation by estate companies, which often becomes a problem in the NES scheme because the PMU scheme supports smallholders without the involvement of the companies, meaning that 'well thought-out' and 'unhurried' development, which is known as 'moderate industrialization,' can be considered to be one of the alternative development options. This moderate industrialization might be realized by establishing small, dispersed but modern high-yield smallholdings through the PMU scheme that at the same time maintains traditional forestry practices and diverse local livelihoods, such as swidden, orchard and rubber agroforest maintenance. The PMU scheme applied at the national level, however, is facing difficulties that stem from its institutional shortfalls, such as high input/investment and the lack of an effective credit repayment system and land tenure uncertainty in local communities. This chapter suggests some possibilities based on this case study to cope with these problems through a more flexible use of the PMU scheme. There is an urgent need to develop economically attractive but ecologically and socially sound smallholder support institutions that can alleviate the strong expansion pressure of companies that lead to large-scale oil palm monoculture and marginalization of local people through land expropriation.

Chapter 7 - The actual tendency of conventional agriculture to grow crops in monoculture to meet the demands of both bioenergy and food production can result in crop yield decline or soil degradation and increased use of synthetic fertilizers. Rhizospheric and soil microorganisms can play a crucial role in plant growth and plant health, contributing to agroecosystem services in many ways. Most crops and terrestrial plants associated with mycorrhizas can lead to better plant growth under stressful conditions through symbiosis. Various arbuscular mycorrhizal fungi (AMF) structures can be used to estimate AMF abundance and diversity in crop fields. The abundance and diversity of AMF are expected to be lower in monocultures. With the increasing effects of global change on field crops, the associated plant-soil microbial communities have been more studied and their beneficial function more recognized for plant success. The lack of long term field studies fail in giving results of the complex cropping systems. Better knowledge of the wide variation in AMF ecology is important to understand and better manage agroecosystems. This chapter explores the current available information relevant to benefits/drawbacks of AMF symbioses in monocultures. Possible roles of fungal structures and Glomalin are also discussed.

In: Monoculture Farming
Editors: Tapan Kumar Nath and Patrick O'Reilly
ISBN: 978-1-63485-166-4

Chapter 1

COMPATIBILITY OF POLICY AND INSTITUTIONAL REGIMES FOR RUBBER MONOCULTURE PLANTATION DEVELOPMENT IN DIVERSE SOCIAL AND AGRO-ECOLOGICAL ENVIRONMENTS: A STUDY IN THE INDIAN CONTEXT

P.K. Viswanathan **and Indraneel Bhowmik***
Gujarat Institute of Development Research, Ahmedabad, India and Tripura University, Tripura, India

ABSTRACT

Policies and institutional interventions for agricultural development in the backward Northeastern region continue to be a prerogative for the national government and the respective state governments in India since long. Various national and state agricultural development agencies, especially the commodity boards have been constantly engaged in the development of commercial agriculture in the region. Among the commodity boards, the interventions by the Rubber Board have been quite significant in terms of social and economic impacts and the entire NER is emerging as the 'Hub of rubber production' in the country accounting for 18 percent of the total area and 6 percent of the total rubber production during 2013. The process of rubber expansion has been quite dramatic, even surpassing the growth observed in the traditional rubber growing states in south India, including Kerala, Tamilnadu and Karnataka. The phenomenal growth of rubber plantation areas in the NE region is also a matter of serious concern, as rubber is heavily promoted as a monoculture system in the otherwise diverse agro-ecological environments of the NER. Rubber expansion in the NER has been justified on the grounds of: (a) ever increasing domestic demand for natural rubber from the manufacturing sector (dominated by tyre industry); and (b) the saturation of agro-climatically suitable lands in the traditional regions, especially, Kerala. Moreover, promotion of rubber cultivation in the NER has been considered to have greater social impacts in terms of rehabilitating the erstwhile shifting cultivators in the region and thereby leading to their social and

* Corresponding author: P.K. Viswanathan (pkviswam@gmail.com).

economic empowerment. However, it emerges that the specific policy and institutional interventions have been oriented towards promoting rubber as a monoculture system in the NE region, which seem to pose potential threats to the co-existing land use and resource management as well as integrated livelihood systems followed by the rural households, dominated by tribal communities in the region.

In this backdrop, the Chapter makes a critical analysis of the compatibility of the policies and institutional interventions historically followed by the Indian Rubber Board in the wide-scale promotion of rubber monoculture in the North Eastern region in India in a comparative perspective of the traditional rubber growing regions in South India, with particular reference to Kerala. If we examine the trajectory of development of rubber plantations in Kerala under the institutional interventions spearheaded by the Rubber Board, it emerges that the Board had promoted a system of rubber production that was highly oriented towards monoculture without considering the crop promotion from an integrated farming system perspective. The chapter further argues that given the agro-ecological diversity and the specific socioeconomic, ethnic and institutional settings as well as the pattern of livelihoods followed, the policies and institutional interventions for rubber development in the NER invariably requires an integrated and holistic approach, so as to minimise the damages caused to the fragile agro-ecosystems of the region and thereby not disturbing the inland fishery, livestock and forestry integrated livelihood systems in the region. Replication of the rubber monoculture as had been widely promoted in the traditional regions, especially, Kerala to the NER, can be a cause of conflict with the pre-existing as well as coexisting agricultural production (including food crops) practices/ farm integrated livelihood systems. In the emerging context, the institutional makeover, including infrastructure support of the Rubber Board in the region also require major restructuring to evolve an integrated approach towards rubber development along with promotion of other livelihood and rubber integrated agro-forestry systems.

THE RATIONALE

India's north eastern region (NER) comprising the seven sisterly states, *viz.*, Arunachal Pradesh, Assam, Manipur, Meghalaya, Mizoram, Nagaland and Tripura and the Himalayan state of Sikkim (Figure 1) accounts for about 8% of the geographical area and 4% of the country'spopulation. NER is connected to East India via a narrow corridor squeezed between independent nations of Bhutan, Myanmar and Bangladesh. The population density of the region (174 persons/sq km) is less than half of the national average (368 persons/sq km), though it varies from a high of 397 persons/sq km in Assam to a low of 17 persons/sq km in Arunachal Pradesh.

The extent of tribal population (27%) in the region is three times higher than the national average (8.6%), with Mizoram (94.4%), Nagaland (86.5%), Meghalaya (86%) and Arunachal Pradesh (69%) exhibiting the highest trend. The region is nevertheless, quite diverse in terms of languages, cultural ethos and pursuit of livelihoods.

Owing to the unique socio-economic, demographic as well as agro-ecological specificities, the trajectory of agricultural development and transformation in the NER has always been a matter of serious contention among academic, policy as well as environmental activist circles. Yet, despite such diverse agriculture and the abundant natural resources, the fact remains that larger part of the region still remains to be backward in terms of development and commercialisation of agriculture in comparison to the states in India's mainland. People are predominantly agrarian and the state of agriculture mostly remains primitive and under developed (Appendix 1). It may even be argued that the economies of the states in the region are yet to experience the kind of structural transformation taken place in a typical developed state of mainland India. Farmers are mostly marginal and rainfed rice (including upland rice) is the staple grain grown with minimum usage of fertilizers and pesticides. *Jhuming* (shifting cultivation) has been the traditional economic practice for a large number of communities in the region[1]though its intensity had declined in recent decades. Further, the overall economic development of the region has been afflicted by problems of growing socio-political unrest, unemployment, food deficit, ethnic conflicts and human rights issues, drug trafficking, immigration, militant turmoil and insurgency (Fernandes, 2004; Shimray 2004). The landlocked nature of the region has imposed further constraints on its economic prospects (Prabhakara, 2004).

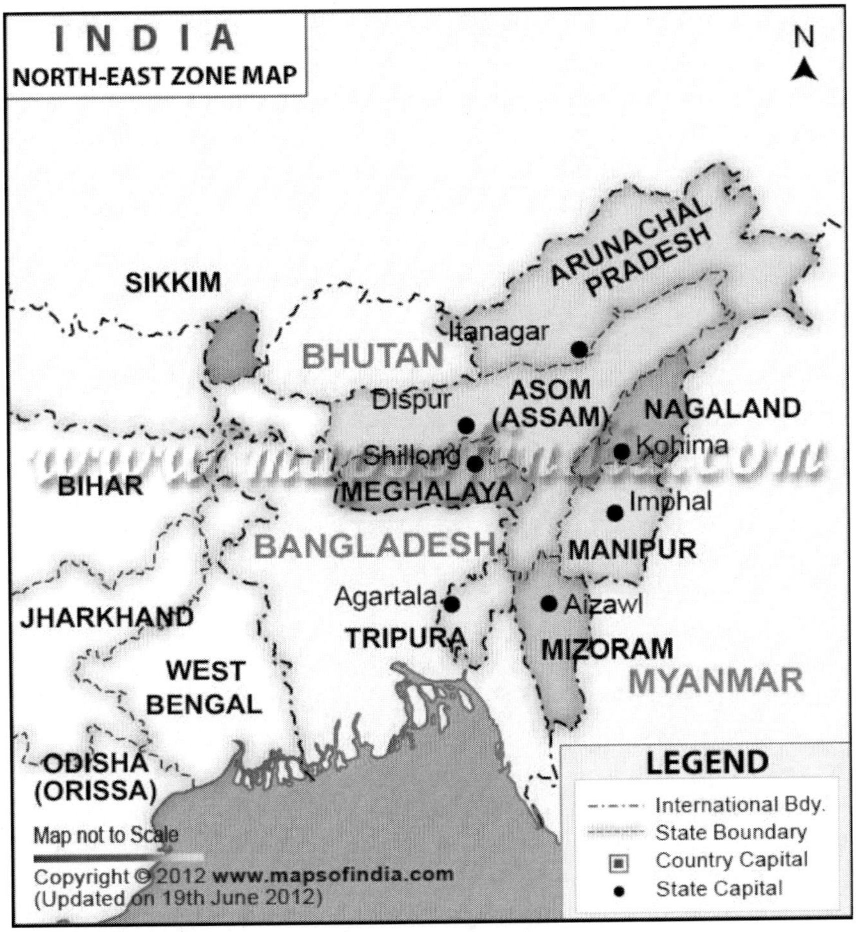

Source: www.mapsofindia.com.

Figure 1. Map showing the geographical boundaries of the North Eastern States.

[1]The extent of shifting cultivation practiced in NE region was reported to be about 14660 sq km (which is about 6% of the total land area of the region) per annum and there are about 4.5 lakh families reportedly engaged in shifting cultivation. As per the Compendium of Environmental Statistics 2008-09 & 2012 prepared by the Ministry of Statistics and Planning (MOSPI), the total area under shifting cultivation in the NE states is 34560 sq km.

Both national and the respective state governments were seriously concerned about the sorry state of affairs of the region and efforts and interventions were on ever since the 1970s for integrating the almost isolated NER with the mainland. Many of these interventions focused on various social and economic upliftment programmes, including development and modernization of agriculture, along with transformation of shifting cultivation with a thrust on horticulture and cash crop promotions. Plantation/ cash crops, such as coffee, tea, rubber, cashew and spices crops, besides horticultural crops such as pineapple and citrus were also promoted under various government programmes as alternatives to shifting cultivation[2].

Development and expansion of cash crops are taking place in the region at the behest of the Commodity Boards, *viz.*, Coffee Board, Tea Board, Rubber Board and Spices Board, who take a lead under the centrally sponsored programmes. Such agency-led crop development programmes were initially intended to make a demonstration effect and were largely planned and maintained by the departments on individual farm lands made available by the farmers or common pool lands owned and operated by village (tribal) communities. Initially, the involvement of the villagers was limited in terms of providing their labour services for establishment and maintenance of the plantations during the immaturity (gestation) period. Once attained productivity, these plantations were handed over to the communities to manage and take the benefits from the sale of the main produce and the ancillary products. To encourage subsequent adoption and expansion of such cash crops, subsidies, extension and market services are provided to the farmers through specific schemes promoted by the respective crop promotional agencies/ commodity boards (Viswanathan, 2006; Choudhury, 2012).

Despite the favourable policies and timely interventions made by the crop development agencies, the success of such programmes was not really encouraging until recently with the exceptions of rubber and tea. Factors such as the lack of familiarity with the crops and their management, difficulty in sourcing seed and saplings as well as inadequate access to marketing and technical backstopping make the farmers hesitant to adopt some of these commercial crops. Reportedly, farmers seem to be less enthusiastic to grow even horticultural crops such as pineapple and other fruits because of the high perishability as well as the persistent marketing problems and the lack of processing facilities, available for these crops. Moreover, it is also reported that farmers who grew many of such crops had to make distress sales in view of the marketing problems and lack of storage facilities (Choudhury, 2012). The much more striking and critical factors that perhaps hinder the wide-scale adoption of many of the cash crops, including tea, coffee, rubber and spices are the disparate gestation periods during which there are hardly any economic returns other than the wage earnings received by the farmers while planting the crops. The virtual absence or lack of access to markets or processing facilities and the price risks involved are also critical issues that would have resulted in the lukewarm adoption of these crops[3].

It may be noted that in a context when many of the commercial crops found very little adoption in NER, the case of expansion of rubber cultivation was unique. Though rubber was

[2]Schemes were also launched to encourage terrace cultivation with the twin objectives of expansion of wet rice cultivation (thus reducing the dependency on shifting cultivation for rice) and promoting soil erosion control (Choudhury, 2012).

[3] It is also reported that the restrictions put by the Coffee Board till the late 1990s on the sale of coffee in the open market (coffee could be sold only to the Coffee Board) frustrated the farmers, who abandoned the plantations. In many places, plantations of coffee, cashew and other cash crops were cut down consequent to the poor market linkages and farmers reluctantly reverted back to shifting cultivation (Choudhury, 2012).

first introduced in the Cachar district in Assam by the British as early as in 1913 (Guha, 1991), it was only since the late 1980s that the first generation of rubber plantations were established on a commercial scale in the NER by the Indian Rubber Board (Viswanathan, 2006). The period since then witnessed a rapid and tremendous expansion of monoculture rubber plantations in the NER especially in the states of Tripura, Assam and Meghalaya. Unlike the other commercial crops, including plantation crops of tea and coffee, the growth in adoption of rubber among the tribal economies of the NER has been quite remarkable over the past two and a half decades, as indicated by the emergence of Tripura as the second largest rubber growing state in India after Kerala in south India. Interestingly, by now, while Tripura has earned the distinction of the '*Second Rubber Capital of India*', the entire NER is emerging as the '*Hub of rubber production*' in the country.

However, the question of compatibility of these policies and institutional interventions for rubber development (that were primarily evolved for the traditional rubber growing regions of south India) in the context of the North Eastern Region (NER) needs a critical assessment. This is because, invariably monoculture rubber expansion get a major boost in the NER in the pretext of rubber development, due to the increasing scope and growth potential for rubber in the region as a strategic product facilitating socio-economic empowerment of the tribal communities in particular, along with integration of the region with the global market.

OBJECTIVES, METHODS AND CONCEPTUAL FRAMEWORK

Against this backdrop, this chapter makes a critical analysis of the issue of compatibility of the policies and institutional interventions for rubber development in the NE states from a comparative perspective of Kerala. The specific objectives of the chapter are to:

1. Provide a comparative perspective on the policy and institutional interventions for the development of rubber plantations in the traditional regions, especially, Kerala *vis a vis* the NE region;
2. Examine the overall economic and social development outcomes realized by the tribal communities in the NER emerging from the existing institutional interventions; and
3. Bring out the major policy as well as institutional development challenges that constrain the development of rubber integrated farming and livelihood systems in the NE context.

We follow a mixed methodological approach in this chapter by integrating the major components, such as review of development interventions, secondary data analysis, primary survey among the beneficiaries of rubber development in the select locations of Kerala and the three major NE states, *viz.*, Tripura, Meghalaya and Assam, which together accounted for almost 81% of total rubber planted area and 91% of total rubber production in India during 2013.

The chapter is organized into five sections, including this introductory. Section 2 provides an overview of the institutional models of plantation development with particular focus on monoculture rubber development in the South and Southeast Asian countries, which has several commonalities with the rubber and related plantation development programmes in India. The section then provides a brief overview of the policy and institutional interventions for rubber development in India from in a comparative perspective of Kerala and NER. Section 3 discusses the major outcomes and impacts of the rubber development programmes in the NER with specific reference to the three major NE states, *viz.*, Tripura, Assam and Meghalaya, which together accounted for about 88% of the rubber planted area and 94% of production in the region during 2013. The section also reviews the overall impacts of the institutional interventions in plantation development in the region with focus on the economic and financial strengthening and empowerment of the regional economies. Section 4 discusses the major challenges and issues surfacing the compatibility and sustainability of the institutional model of rubber development in the NER. Section 5 concludes the chapter highlighting the policy and institutional imperatives emerging from the study.

Following the Institutional Analysis and Development (IAD) framework as proposed by Ostrom (1990; 2005; 2007), this chapter uses a broader conceptual definition to the term *institutions*. Accordingly, we define institutions as a set of prescriptions that include rules, norms, and shared strategies (Ostrom, 2005: 3). Institutions are further delineated as being formal or informal; the former characterized as rules-in-form and the latter as rules-in-use. In the specific context of rubber, we consider all policies and interventions being adopted by the institutional agency, *viz.*, Rubber Board, for the overall development of the rubber sector as well as socio-economic upliftment of the beneficiary communities (mostly tribal) in the North Eastern states in particular. For conceptual clarity, we consider the important policies and interventions undertaken by the Rubber Board as cutting across the technological, institutional and organizational domains of rubber production and management in India. Needless to say that all these policies and interventions have been targeted at: (a) strengthening the capacity of the domestic rubber production sector; (b) enhancing its trade competitiveness; and (c) sustaining the livelihoods of the small and marginal producers and the dependent labour communities.

INSTITUTIONAL MODELS OF PLANTATIONS DEVELOPMENT: THE CASE OF NATURAL RUBBER

The history of evolution of plantation development in the South and Southeast Asian countries is replete with the emergence and continued existence of several institutional models in the case of the major plantation products, *viz.*, tea, rubber, oil palm, coffee, spices crops (pepper, cocoa, and cardamom), sugarcane, cassava, banana, etc. Among these plantation commodities, the case of rubber is distinct in terms of three specific models, *viz.*, (a) the 'large scale monoculture plantation model' of the colonial vintage (late 19th to mid-20th centuries); (b) the dominant small and medium farmer (SMF) plantation model (evolved during the intervening period between 1940s through 1990s); and (c) agri-business oriented plantation models, being currently (early 21st century) promoted in the emerging economies of Myanmar, Cambodia and Lao PDR facilitated by the inflow of financial capital owned by the foreign firms located in China, Vietnam and Thailand (Barlow, 1997; George et al., 1988; Lipton, 2009; Hayami, 2010;Viswanathan, 2006; Byerlee, 2014; Byerlee *et al.*, 2014).

Interestingly, the three rubber plantation models listed, have several distinctions with respect to their developmental outcomes as well as the interface with the economy, society, environment and ecosystems prevailing in the regions/ areas where monoculture rubber plantations have been established. Moreover, though the models broadly depict the structure and organisation of production, they may vary in terms of production relations, processing and product marketing arrangements, composition of the value chain, extent of value addition as well as the distribution or sharing of gains (farm business income) among the various actors in the value chain, especially, the small and marginal producers and the workers in the case of the SMF model.

The characteristic features, the development outcomes as well as the socio-economic and environmental (ecological) interface of these three institutional models of rubber plantations are presented in Table 1. It may be observed that amongst the three plantation models, the small and marginal farmer (SMF) model had been widely adopted by most rubber producing countries *viz.*, Malaysia, Thailand, Indonesia, India, Bangladesh, Vietnam, Myanmar and Cambodia. In fact, smallholder sector accounts for about 85-90% of the total area and production of rubber in these countries (Byerlee, 2014) and the plantations are distinct in terms of their monoculture orientation with few exceptions as in the case of Jungle rubber in Indonesia (Joshi *et al.*, 2002) and rubber integrated farm livelihood systems in Thailand (Somboonsuke, *et al.*, 2001; Viswanathan, 2006). It is also important to note that the pathway of rubber development in these countries, including India had been greatly influenced by the SMF model (SMFRDM) with orientation towards monoculture, due mainly to the smallholder friendly policy and institutional interventions by the national and provincial governments in these countries over time. Table 2 presents the status of monoculture rubber development in the Asiatic countries and their relative share in the global rubber production.

Table 1. Features, development outcomes and socio-economic and environmental (ecological) interface of institutional models of rubber plantation development

Distinct features	Country/ Regional context	Promotional agencies: (development, research, extension and marketing)	Development outcomes	Socio-economic and environmental (ecological) interface
1. Large Scale Plantation Rubber Development model (LSPRDM): [Since late 1800 continue – but has been on the decline in countries, such as India]				
Plantations admeasuring more than 20-50 hectares (average size ranging from 200 ha-2000 ha)	Malaysia, India, Indonesia, Sri Lanka, Thailand, China, Vietnam	Colonial governments/ FELDA/ RISDA/ MRB/ RRIM/ IRB/ RRII/ ORRAF/ RRIT/ RRISL/ VRA	Integration of production,primary processing, product development, whole-sale marketing and exports activities	(a) Employment generation (b) Economic growth (c) Monoculture orientation/ carbon sequestration (d) Process and product waste generation, (e) Pollution problems- dedicated environmental management systems in place

Table 1. (Continued)

Distinct features	Country/ Regional context	Promotional agencies: (development, research, extension and marketing)	Development outcomes	Socio-economic and environmental (ecological) interface
2. Small and Medium Farmer Rubber Development model (SMFRDM) [dominant model since 1940s – going to stay in the future with few exceptions]				
Plantations are small and marginal holdings below 20 ha (majority being below 2 ha- average of 1 ha & below in Indian states, including NER	Malaysia, India, Thailand, China, Indonesia, Bangladesh, Lao PDR, Cambodia, Myanmar, Vietnam	Government agencies/ State Farms in China [70-73% as of 2004- farmers work as production workers on contracted farms]/Commodity Boards (MRB/ RB/CRRI/ ORRAF)/Rubber Producer Co-operatives/ Block or Group Planting Units	Individual and Collective (co-operative) systems of production, processing and marketing	(a) Most inclusive model with provision of employment, livelihood support to the plantation based communities directly and indirectly; (b) Though environmentally sound ways of production and processing are available, compliance is often found to be missing; (c) Monoculture orientation is a threat to the pre and co-existing land use systems, ecological and hydrological regimes; (d) But, there are sustainable smallholder agro-forestry models as exist in Indonesia (Jungle Rubber), integration of food crop production and other livelihood (fishery, livestock, fruit tree production) systems in Thailand- rubber production in the NER offers a major potential in this case (e) issues in sharing of carbon credit benefits
3. Agri-business Rubber Development model (ABRDM) [newly developed and yet emerging model]				
Plantations developed by foreign owned companies/ FDI through land concessions obtained from the donor countries – average size of land concessions range between 5000 – 10000 ha admeasuring more than 20-50 hectares (average size ranging from 200 ha-2000 ha) [Viswanathan, 2013]	Lao PDR, Cambodia and Myanmar account for almost 60% (1.74 million ha) of the total global area of concessions under rubber (2.95 million ha) [Byerlee et al., 2014]	Private investors/ firms/ nationals from China, Vietnam and Thailand with loan support from International financial firms	Bulk of the rubber output marketed to export oriented firms in China, Vietnam and Thailand after primary processing	(a) Least inclusive in terms of growth in the countries offering the land concessions, except the employment benefits; (b) The surge in land concessions has led to issues of negative social and environmental impacts. (c) Poorly defined property rights lead to many claims of overlapping land rights and displacement of existing users, displacement of livestock communities (the case of Lao PDR) [Viswanathan, 2013]; (d) Land issues often lead to conflict (in Cambodia land issues were surfaced in 2013 elections that even sharply reduced the support to the long ruling party [Byerlee *et al.*, 2014].

Notes: MRB- Malaysian Rubber Board; ORRAF - Office of Rubber Replanting Aid Fund; IRB – Indian Rubber Board; RRII – Rubber Research Institute of India; RRISL – Rubber Research Institute of Sri Lanka; RRIT – Rubber Research Institute of Thailand; VRA – Vietnam Rubber Association. CRRI – Cambodia Rubber Research Institute; RRIM- Rubber Research Institute of Malaysia; FELDA – Federal Land Development Authority (Malaysia); RISDA - Rubber Industry Smallholders' Development Authority (Malaysia).

Source: Authors' compilation based on various rubber plantation related studies, including Viswanathan, 2006; 2008 & 2013; Byerlee, *et al.*, 2014.

Table 2. Status of major countries in area and production of natural rubber, 2013

Countries	Harvested area ('000 Ha)	% share	Production ('000 Tonnes)	% share
1. Indonesia	3556	34.47	3108	25.97
2. Thailand	2421	23.47	3863	32.28
3. Malaysia	1057	10.25	826	6.91
4. China	686	6.65	865	7.23
5. Vietnam	548	5.31	949	7.93
6. India	442	4.28	900	7.52
7. Myanmar	204	1.98	148	1.24
8. Philippines	185	1.79	111	0.93
9. Sri Lanka	136	1.32	130	1.09
All Asian Countries	9335	90.50	10950	91.51
World	10316	100.00	11966	100.00

Source: Compiled from FAOSTAT database.

Institutional Interventions for Rubber Development in India: A Comparative Perspective of Kerala and NER

If we examine the trajectory of development of rubber plantations in the traditional region of Kerala *vis a vis* the non-traditional regions of NE states, it emerges that the institutional model of rubber development as evolved for the traditional regions in a particular historical context, has been replicated to the non-traditional regions, especially, the NE states. A critical assessment of the institutional model of rubber development would reveal that the system of rubber production developed in the traditional regions was highly oriented towards development of monoculture regardless of the potentials of an integrated agriculture system involving multiple crops and livelihood practices. As a result of the intensive rubber monoculture plantation development activities being pursued by the Rubber Board, it may be seen that Kerala, which is a severely land constrained state, had lost much of its pre-existed integrated land use and farm management practices to rubber monoculture. Unlike other commodity boards and agencies, the institutional interventions by the Rubber Board have earned wider acclaim and acceptance in the traditional regions of Kerala and other states in view of their intensive crop promotional strategies and farmer outreach and support programmes. Apparently, the more elaborate and all-encompassing interventions by the Rubber Board rendered it prominent over other crop promotional agencies, which in turn, have also made the other commercial crop- based institutional agencies and interventions (like the Coconut Development Board, Coffee Board, Spices Board, state government schemes for promotion of paddy, Kerala Horticulture Development Board, etc.) less effective to a large extent in Kerala. Given the specific socio-economic, historic and political factors, the institutional architecture evolved for the promotion of rubber had been quite successful in the traditional regions of Kerala and others, in terms of its perceived targets and goals as well as developmental outcomes. In other words, the policies and institutional interventions had turned out to be highly successful in the specific context of Kerala and its legitimacy had never been contested by any political, developmental, environmental or civil society bodies, for well-known reasons.

In what follows, we present an overview of the specific institutional interventions in the development of the rubber smallholder sector in India under the aegis of the Rubber Board since its formation in 1954 under the Rubber (Production and Marketing) Act 1947.

In India, rubber plantations were first established in the southernmost state of Kerala as early as 1902 by the British colonial government. The period since Independence had witnessed tremendous expansion of rubber plantations in the southern states (traditional regions), dominated by Kerala, Tamilnadu and Karnataka, following the setting up of the Rubber Board in 1954 and the Rubber Research Institute of India in 1955 under the Ministry of Commerce and Industry. The process of development of rubber plantations gathered momentum in the traditional regions in the 1960s through 1980s under this institutional and policy support regime leading to the emergence of a dynamic smallholder sector in the country. Apparently, the growth of the rubber smallholder sector in Kerala, in particular, has been contingent upon the institutional architecture created under the aegis of the Rubber Board and the R&D support system provided by the Rubber Research Institute of India (George, *et al.*, 1988; Viswanathan, and Shivakoti, 2007; 2008).

The policy and institutional architecture comprising an array of smallholder support measures, broadly called as Rubber Plantation Development (RPD) schemes ranging from planting (newplanting and replanting) subsidies to market protection (Box 1), had attracted a large segment of the enterprising native peasantry to take up rubber cultivation as the major source of livelihood.

This was also facilitated by a host of political, socio-economic and institutional factors, including the land reform policy, besides the favourable agro-climatic conditions prevailed in Kerala. Over time, the dynamic smallholder sector has far exceeded the estate sector in area and production of rubber and currently, the share of the smallholdings is as high as 89% in area and 93% in production of rubber (Viswanathan and Shah, 2012; 2013).

While the area expansion under monoculture rubber had relentlessly continued in the traditional regions, especially, Kerala, the need for expanding it onto the non-traditional regions became imperative especially since the late 1980s. This was mainly due to: (a) the ever increasing domestic demand for natural rubber from the manufacturing sector (dominated by tyre industry); and (b) the saturation of agro-climatically suitable lands in the traditional regions, especially, Kerala. In the process of expansion, the North Eastern states have been identified as the potential regions due to the agro-climatic suitability of the region. Hence, efforts had been on, to develop rubber plantations in the NER, which was also legitimized by the national government in view of the perceived developmental goals that the rubber development programmes would bring forth social and economic upliftment and mainstreaming of the tribal communities in the NE region, a vast majority of whom were poor and had been allegedly engaged in shifting cultivation practices.

The current scenario of development of rubber plantations in India is quite revealing in that the share of traditional regions (southern states, dominated by Kerala) had declined from almost 100% during 1960-61 to 82% during 2011-12 (Table 3). On the other hand, the share of the NER (comprising 7 states) had almost doubled from mere 7% to almost 18% during the last two decades. More importantly, there was almost four-fold increase in rubber area in the North Eastern states in absolute terms from 33619 ha (1990-91) to 1,28,470 ha (2011-12) [also see Figure 2].

One of the important outcomes of the institutional interventions in case of rubber in Kerala's agricultural landscape was the emergence of monoculture rubber as the second major crop in the state after coconut in terms of relative share in gross cropped area. In fact, rubber occupied almost 20% of the gross cropped area (the share of coconut being 29%) during 2009-10 (GOK, 2011).

Box 1. Major Policy and Institutional Interventions of Rubber Board in India

Support measures	Description/ scale of activities
1. Newplanting and Replanting subsidies	The scale of assistance provided: INR 25,000 (US$ 373) per ha for traditional regions and INR35,000 (US$ 522) per ha for non-traditional regions, including NER. Subsidy provided in 6 annual installments. Rubber Board also facilitates financial credit from commercial banks.
2. Provision of Input subsidies; Setting up or financial assistance to Rubber Nurseries; Promotion of scientific farming practices and plant protection measures	For purchase of planting materials, fertilizer, material inputs, sprayers, setting up of smoke house, purchase of rubber roller, weed cutter, etc. Growers in non-traditional regions are eligible for reimbursement of cost of planting materials and transportation grant @ INR 4000 (US$ 60) per ha.
3. Formation of Farmer Groups and Empowerment	Technical and financial support for setting up of Rubber Producers' (Growers') Societies (RPS/RGS); Block Planting Units (RPUs); Group Processing Centres (GPCs); Women Self Help Groups (SHGs); Rubber Agro-Management Units (RAMUs), etc.
4. Training in Rubber Growing, Tapping and Processing	Tappers Training Schools (TTS) have been established for imparting training on scientific ways of rubber tapping. Growers from Non-traditional regions are given training in rubber growing, tapping and processing.
5. Primary Processing, Marketing, Product Development and Exports of rubber	Networking of Rubber Producers Societies with the Private Rubber Companies for marketing and trade of processed rubber.
6. Price Stabilization Fund (PSF)	PSF is a Trust Fund under the aegis of the NABARD and aims to provide income support to rubber growers when price of rubber falls below remunerative level.
7. Other schemes of general nature	Award for best RPS; Award for best rubber grower; Commercial Promotion of rubber honey and rubber wood; Promotion of rubber based industries.

Note: 1 US $ = INR 67 (2016)

Source: Compiled from: Rubber Plantation Development Scheme (Phase VI Rules): Newplanting & Replanting Components 2007-08 – 2011-12 [www. rubberboard.org.in].

Table 3. Status of monoculture rubber plantation development in India (Area in ha)

Year	Kerala	Tamilnadu & Karnataka	Southern States	North Eastern States[A]	Other states[B]	All India
1960-61	135809 (94.4)	7915 (5.5)	143724 (99.9)	---	181 (0.13)	143905 (100)
1990-91	407821 (85.8)	31145 (6.6)	438966 (92.4)	33619 (7.1)	2498 (0.53)	475083 (100)
2000-01	474365 (84.3)	38445 (6.8)	512810 (91.1)	46885 (8.3)	2975 (0.53)	562670 (100)
2006-07	502240 (81.6)	45268 (7.4)	547508 (89.0)	64883 (10.6)	2809 (0.46)	615200 (100)
2007-08	512045 (80.6)	48240 (7.6)	560396 (88.2)	71480 (11.2)	3524 (0.55)	635400 (100)
2011-12	539565 (73.4)	61378 (8.4)	600943 (81.8)	128470 (17.5)	5367 (0.73)	734780 (100)

Note: Figures in parentheses are respective shares at the All India level. A- North Eastern states comprise of Assam, Meghalaya, Tripura, Arunachal Pradesh, Nagaland, Sikkim and Manipur. B- Other states include Maharashtra, Orissa, Andhra, and West Bengal.

Source: Viswanathan and Shah, 2013.

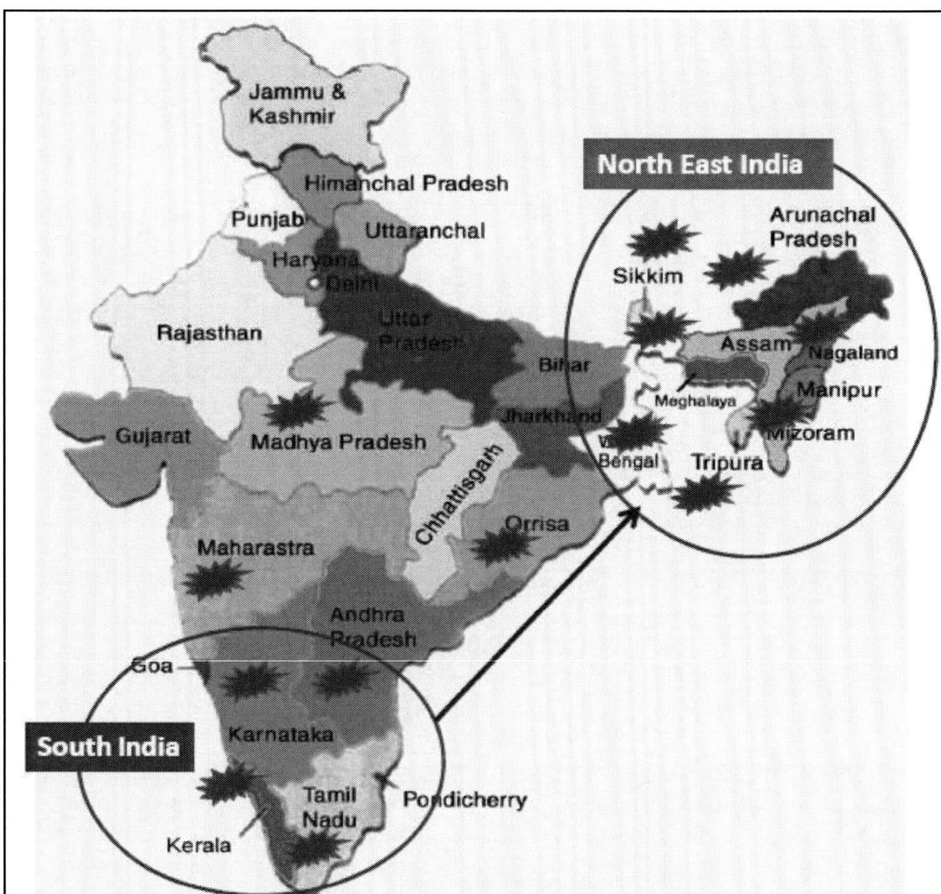

Note: The rubber growing regions are shown in the green shades.

Source: Viswanathan and Shivakoti, 2008.

Figure 2. Map showing monoculture rubber-growing regions across states, including NER in India.

The impacts of the institutional interventions of the Rubber Board have been quite pervasive in Kerala, with the result that the relative share of rubber in area as well as gross value of agricultural output (GVAO) had grown in all the districts (except Alapuzha and Wynad) as evident from Table 4.

Apparently, while the relative share of rubber in gross cropped area seems to have outpaced that of coconut in four districts, its share in gross value of agricultural output had grown over coconut in 11 of the 14 districts (exceptions being Alapuzha, Thrissur and Kozhikode).

Table 4. Relative share of monoculture rubber versus other crops in gross cropped area and the gross value of agricultural output in Kerala, district-wise

District	Share of major crops in Gross Cropped Area (%), 2011-12			Relative share of major crops in Gross value of agri. Output (%)					
				Rice		Coconut		Rubber	
	Rice	Coconut	Rubber	2003-04	2007-08	2003-04	2007-08	2003-04	2007-08
Trivandrum*	1.8	43.9	18.4	1.7	0.7	43.2	26.1	25.0	40.4
Kollam	2.3	35.0	21.4	2.7	0.8	34.4	18.0	30.9	42.9
Pathanamthitta	2.6	16.4	49.0	2.4	0.5	14.3	6.2	61.4	66.8
Alapuzha	32.5	37.0	4.1	21.4	18.9	56.1	35.9	7.6	14.1
Kottayam	5.1	16.1	52.0	2.7	2.1	12.9	6.0	70.8	79.6
Idukki	0.7	6.2	13.8	0.9	0.6	6.3	2.7	23.7	32.7
Ernakulam	7.2	25.8	32.5	7.0	2.1	24.1	12.0	49.5	63.3
Thrissur	16.1	44.3	8.5	13.9	9.5	55.1	35.2	18.0	32.6
Palakkad	30.6	18.8	11.1	27.4	22.4	25.1	12.3	25.8	35.8
Malappuram	4.5	43.6	15.2	4.1	1.8	45.5	24.2	22.2	37.8
Kozhikode	2.0	59.1	9.8	1.1	0.8	65.4	40.9	15.2	33.9
Wynad	6.7	5.8	4.9	6.3	5.3	6.1	3.8	6.6	10.8
Kannur	3.6	36.5	19.9	2.1	1.5	38.9	19.2	30.7	51.1
Kasargode	3.6	37.5	21.1	1.8	1.1	47.7	17.8	19.9	32.8
KERALA	8.7	29.2	19.2	5.9	4.1	32.7	16.2	31.3	45.3

Note: * The name Trivandrum has been changed into Thiruvananthapuram in 1991.
Source: Viswanathan, 2014.

In order to further capture the extent of rubber monoculture *vis a vis* other crops currently being practiced in Kerala, we have used the data source based on the Land Resources Information System (LRIS), which is a web-based Geographic Information System (GIS) launched by the Kerala State Land Use Board (KSLUB) in 2013. Accordingly, this dataset provides the latest land use information for different blocks of seven districts, viz., Kottayam,

Ernakulam, Thrissur, Palakkad, Kannur, Idukki and Thiruvananthapuram[4]. The summary of the cropping pattern based land use data is presented in Table 5.

The cropping pattern in Kerala as presented in Table 5 shows the predominance of rubber monoculture across districts and the blocks and in majority of the cases, the extent of rubber monoculture significantly exceeds the proportion of mixed crops. The nature of mixed crops is such that the farmers tend to grow a variety of horticultural and commercial crops, such as mixed vegetables, coconut, arecanut, banana, pepper, cashew, cardamom, etc, within the tiny plot of land owned by them.

Apparently, Kerala had experienced a significant decline in the size of operational farm land holdings due to the prevalence of inheritance laws leading to sub-division and fragmentation of holdings over time. Interestingly, even with such fragmentation of holdings, the expansion of rubber monoculture had continued unperturbed due to the stringent rubber planting policies promoted by the Rubber Board over time. The changes in cropping pattern in the state also witnessed a tremendous decline in area under food crops, mainly paddy as well as a large-scale conversion of coconut, arecanut and cashew garden lands into monoculture rubber plantations due to a host of price and non-price factors, including the promotional policies extended by the Rubber Board. The paradigm shift in cropping pattern in the state characterised by the emergence of monoculture of rubber along with the decline in food crops, such as rice, tapioca as well as commercial crops, such as pepper and arecanut have also been well captured by several scholars (Viswanathan, 2002; Joseph, 2002). Notably, a study by Joseph (2002)[5] had historically traced the changes in cropping pattern and the emergence of monoculture rubber in the North Kerala villages following the migration of peasants from Central Kerala since the 1940s to the 1980s for expanding rubber and other commercial crops in the north Kerala region. The resultant changes in cropping pattern, leading to the virtual monopoly of rubber in the North Kerala villages are presented in Table 6.

Table 5. Status of development of rubber monoculture and other crops in districts of Kerala, major blocks

District	Blocks	Percentage of area under			Total crop Area (ha)
		Rubber monocrop	Mixed crops	Paddy	
1. Kottayam	Erattupetta	57	25	0	27584
	Kaduthuruthy	39	21	6	16067
	Kanjirappally	58	16	0	35191
	Madappally	34	29	0	10469
	Pampady	64	23	1	20966
	Pallam	26	45	5	10655
	Uzhavoor	62	18	2	22609
	Vazhoor	61	26	1	16526

[4]This data source provides a snapshot of the important landuse categories for different blocks coming under the administrative jurisdiction of various districts of Kerala. Currently, this landuse data is available for seven districts, viz., Thiruvananthapuram, Kottayam, Thrissur, Palakkad, Ernakulam, Kannur, Wynad and Idukki. This dataset has been developed by combining manual data gathered through the Panchayat resource mappings (PRM) and information from satellites (http://kslublris.com/LRIS/Kerala/district.php).

[5]The interesting piece of work by Joseph (2002) examining the interface between migration and changing land use pattern in 26 villages in the three districts of North Kerala, *viz.*, Malappuram, Kannur and Kozhikode, traced the dynamic growth of rubber in the agrarian landscape of the North Kerala region over the past six decades.

Table 5. (Continued)

District	Blocks	Percentage of area under			Total crop Area(ha)
		Rubber monocrop	Mixed crops	Paddy	
2. Palakkad	Alathur	37	19	24	31091
	Palakkad	15	24	23	18320
	Mannarkad	28	21	6	43860
	Sreekrishnapuram	26	29	13	22398
3. Ernakulam	Angamaly	19	18	3	39063
	Koovapady	16	33	8	18026
	Kothamangalam*	13	8	1	91594
	Mulanthuruthy	39	24	6	13696
	Muvattupuzha	47	30	3	21047
	Pampakuda	51	26	6	18814
	Vadavukode	35	31	7	18637
	Vazhakulam	26	34	9	13147
4. Thrissur	Kodakara	14	14	2	47149
	Ollukkara	11	23	2	25016
	Pazhayannur	16	17	11	31712
	Vadakkanchery	10	21	12	21824
5. Thiruvananthapuram	Chirayinkizhu	20	32	6	7107
	Kilimanoor	29	19	9	18013
	Nedumangad	28	29	1	12228
	Nemom	15	34	3	11890
	Perumkadavila	29	19	3	25952
	Vamanapuram	17	12	3	43578
	Vellanadu	15	9	5	40328
6. Idukki	Thodupuzha	49	20	1	13432
	Thodupuzha municipality	40	31	2	2794
7. Kannur	Irikkur	31	45	0	29376
	Iritty	17	69	0	13634
	Koothuparamba	18	9	3	5941
	Payyannur	25	51	4	24875
	Peravoor	40	46	3	21410
	Thaliparambu	46	27	2	27778
	Average	31	26	5	23943
	Coefficient of Variation (%)	50.5	46.5	110.4	64.8

Note: * Kothamangalam block reports forest area covered in 69% of the total area.
Source: Compiled from LRIS data set, Kerala Land Use Board.

Thus the expansion and spread of monoculture rubber in any village of Kerala had been quite dramatic and all-encompassing and this competitive monocrop had gradually replaced vast tracts of the other agricultural crops that either had faced low prices or high costs of farm management, including labour costs. A recent study by the author (Viswanathan, 2015) exploring the impact of peasant migration and the resultant changes in cropping pattern in few selected villages of Taliparamba block in North Kerala further brings out the expansion of rubber monoculture as evident from Figure 3. As notable from the Figure 3, the rubber monoculture (as indicated by the blue patches) is wide-spread across the villages. The other land uses include crops, such as rice (paddy), coconut, arecanut, pepper, mixed crops and forests.

Table 6. Area grown with different crops by early migrants in villages in North Kerala (%)

Crops	Before 1940	1941-50	1951-60	1961-70	1971-80	1997	2012$
1. Rice	23.40	13.46	9.63	8.77	2.19	6.91	2.58
2. Tapioca	14.89	12.01	10.84	10.01	6.76	0.68	1.23
3. Coconut	4.25	14.30	21.12	20.00	12.19	22.43	28.70
4. Arecanut	8.51	4.64	12.40	10.27	13.12	11.36	6.22
5. Rubber	---	0.96	2.45	4.25	22.96	32.48	48.14
6. Pepper	---	6.36	8.08	11.22	14.34	5.71	3.19
7. Others*	48.95	48.29	35.50	35.49	28.46	20.43	9.94
Total area (acres)	47.0	1141.0	1308.0	2456.5	2730.5	3254.6	3468.6

Note: * Others include ginger, coffee, cocoa, lemongrass, cashew, vegetables, etc.
Source: Compiled from Joseph, 2002; $- Kerala Land Use Board Database (compiled by the author).

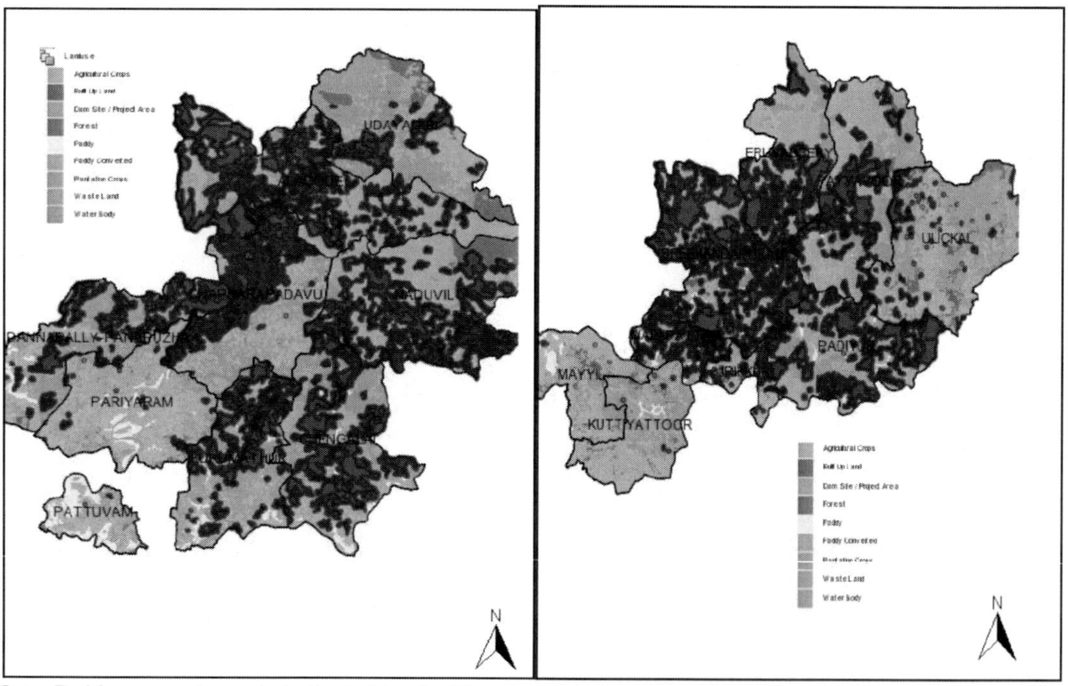

Note: Rubber spread area is indicated by the blue colour across the villages.
Source: Maps generated by the author from the LRIS dataset, Kerala Land Use Board.

Figure 3. Spread of rubber area in the selected villages in Taliparamba block, North Kerala.

POLICY AND INSTITUTIONAL INTERVENTIONS IN RUBBER DEVELOPMENT AND OUTCOMES IN INDIA'S NORTH EASTERN REGION

A closer look at the latest data on rubber planted area and rubber production as presented in Table 7 reveals that the NER accounts for about 19% of the planted area in India, while the contribution of the region to rubber production at the national level is only 6%. This sharp contrast in area and production shares of NER depicts that a large chunk of the rubber plantations are new plantations that have been established during the last 5-6 years. This also suggests that the share of the NER in production would be more than doubled during the next 5-6 years once these young plantations attain tappability. Based on the emerging scenario, it may also be noted that the country's future prospects in rubber production would largely be determined by the expansion of rubber in the NER.

Table 7. State-wise total area and production of rubber monoculture in India, 2012-13

State/ Region	Total rubber area (ha)	(%)share	Total rubber Production (tonnes)	(%)share
1. Kerala	545030	71.95	800050	87.56
2. Tamilnadu	20770	2.74	25350	2.77
3. Karnataka	44900	5.93	31250	3.42
4. Tripura	67730	8.94	33220	3.64
5. Assam	43335	5.72	11740	1.28
6. Meghalaya	12865	1.70	7110	0.78
7. Rest of India	22890	3.02	4980	0.55
All India	757520	100.0	913700	100.0
NE Region	141035	18.62	55280	6.05

Source: Rubber Board (2013): Rubber Statistical News, 72 (6) [November].

In what follows, we make a brief overview of the policy and institutional interventions in rubber development by the Rubber Board and other institutional agencies, including the state agencies in the context of Tripura, Meghalaya and Assam, which together account for 88% of total rubber area and 94% of total rubber production in the North Eastern region (Table 7).

Policy and Institutional Interventions for Rubber Development in Tripura

Tripura leads among the non-traditional rubber growing regions of India. Initiated by the state forest department (SFD) fifty years back as part of afforestation, rubber plantations have now emerged as an engine of growth for the state contributing around Rs. 600 crores per annum which is approximately 3 to 3.5 percent of the Gross State Domestic Product (GSDP) of Tripura. The growth of commercial rubber plantations in Tripura was initially at the behest of the state initiatives. In a historical perspective, the state Forest Department introduced rubber trees in Tripura in 1963. Encouraged by the success of the experimental plantations, the state forest department (SFD) sought to expand plantation cover. Seeing the greater potential for extension, the Rubber Board set up its one man field office in 1967. Nonetheless, being a new commercial crop hitherto unfamiliar in Tripura, rubber was subjected to diverse tests from various quarters of the state and the society. Often, it was abandoned due to misunderstanding and rejection. Hence, though extension activities began at an increasing rate, it was mostly confined to the public sector.

The state government realized that rubber plantations apart from being a source of revenue also had the potential as a viable means for resettlement of the landless shifting cultivators. The Tripura Forest Development & Plantation Corporation (TFDPC) was formed in 1976-77 and was entrusted with the management and operational activity of the existing plantations of the state forest department. Alongside, the corporation was assigned to undertake the rehabilitation of the landless jhum cultivators. The first rubber based rehabilitation package, undertaken by TFDPC, came up in 1976 at Warrangbari, West Tripura. The programme offered a maximum of 1.5 ha of rubber plantation with the owner himself providing the labour. Buoyed at the growing stature of NR in Tripura, the Rubber Board set up its regional office in 1979 at Agartala and subsequently Nucleus Rubber Estate and Training Centre (NRETC) was established in 1984 with the Jt. Rubber Production Commissioner, Rubber Board as the administrative head under the scheme of Accelerated Rubber Development (ARD) in the North-East. Subsequently Regional offices at Udaipur (in South Tripura district) and Dharmanagar (in North Tripura district) were set up in 1988 and 1994 respectively (Paribalan, 2006).

The policy of rubber development and extension activity of the Rubber Board comprising of free distribution of planting materials (i.e., polybag & budded stumps), provisions for maintenance cost, fencing materials and technical advice saw the emergence of increasing number of private plantations also. Moreover, with the Cash Subsidy Scheme introduced by the Rubber Board in 1980, augmentation of new rubber plantations occurred in Tripura. This scheme of financial subsidy and the success of the rehabilitation programme of the TFDPC encouraged the state government to establish the Tripura Rehabilitation & Plantation Corporation (TRPC) in 1983 with the specific objective of settlement of landless tribal people for reclamation of lands under Jhum cultivation.

Subsequently, the "Tripura Block Plantation Project" was introduced in 1992-93 for the settlement of tribal families in collaboration with the Rubber Board and the Department of Tribal Welfare, Govt. of Tripura with sponsorship from the World Bank. Two more agencies to implement rubber plantations- the office of the Tripura Tribal Areas Autonomous District Council (TTAADC) and the office of the Sub-divisional Magistrates (SDM) were set up to promote rubber based development and rehabilitation scheme from 1998.

On the other hand, the Rubber Board with its increased official reachout in the state pursued its policy of extension among the private growers. The economic reforms and the succeeding industrial spurt in the early 1990s resulted in a spectacular rise in the prices of rubber particularly since 1994 and had attracted a lot of private initiatives. But, the subsequent depression in the international rubber market in the late 1990s had a toll on the rubber economy of Tripura and the growth was unstable. However, the international NR sector indicated a revival since 2001, which naturally resulted in renewed interest in rubber plantations among the private sector investors.

The state government realizing the potential of rubber as an instrument for development, in 2006, constituted "the Tripura Rubber Mission in the Forest Department with the objectives to bring a total of 85094 ha land under rubber cultivation in the next 20 years"

(GOT, 2006). The new areas were supposed to include both non-forest as well as degraded forestland spread along the international border with Bangladesh and by the side of the national and state highways (Bahuguna, 2005).

Rubber based rehabilitation in Tripura has been a success story as is evident from the numerous cases and reports about the improved life and livelihood of erstwhile shifting cultivators (Pereira, 2009; Chakraborty, 2012). Table 8 shows that the three major agencies adopting the rehabilitation model have helped in the resettlement of more than 20000 tribal beneficiaries and these rehabilitated rubber growers, in unison, currently holds control over almost one-third of the total rubber area in Tripura. Thus, it may be observed that the rehabilitation model adopted by the major public agencies of the state had created the foundation for the rubber success story that the state showcases.

Table 8. Rubber based rehabilitation by major development agencies in Tripura

Agency	Total Beneficiaries (Numbers)	Total Area (ha)	(%) share
1. TFDPC	3250	3268	16.6
2. TRPC	7285	6600	33.5
3. Rubber Board	9778	9845	49.9
Total (3 agencies)	20313	19713	100.0

Notes: There are other smaller agencies like TTAADC and office of the SDM pursuing rubber based rehabilitation, but their contribution is much lesser

Source: Annon. (2012a), Annon. (2012b), Rubber Board (2013).

The interventions by the Rubber Board was unique in terms of introducing a block planting scheme (BPS) in Tripura in 1992 in three different units/blocks with a total area of 113.99 ha (Rubber Board, 1992). Under the BPS, a compact land (block) owned by scheduled caste and scheduled tribe (SC & ST) households is identified and the plantation is raised by engaging family labour as wage earners. The important criteria for raising a block planting unit (BPU) are: (i) minimum extent of land available in each colony shall be around 50 ha.; (ii) the land shall be in contiguous plots; (iii) tribal beneficiaries who own land should hand over the land to the Rubber Board for a period of seven years; (iv) assurance from beneficiaries for the engagement of family labour during the immature phase; (v) the beneficiary should have the clear title to the land; (vi) the extent of land ranges from around 1 ha to 2 ha; (vii) formation of RPS after BPUs start yielding rubber; (viii) retention of the plantation by Rubber Board for two more years after opening the trees for tapping in order for stabilizing the harvesting, processing and marketing procedures before handing over the plots back to the beneficiaries; (ix) the financial expenditure will be borne by the State Government and the Rubber Board, while beneficiary contributions are in the form of family labour (Rubber Board, 1992; 1997; 2005). The compulsory engagement of family labour as wage earners ensures employment and income to the beneficiary households during the immature phase (Viswanathan and George, 2005; Joseph et al., 2010). Figure 4 shows the increase in area of rubber developed under the block planting scheme in Tripura. However, the area planted under the BPS constituted hardly 9% of the total rubber planted area during 2006-07.

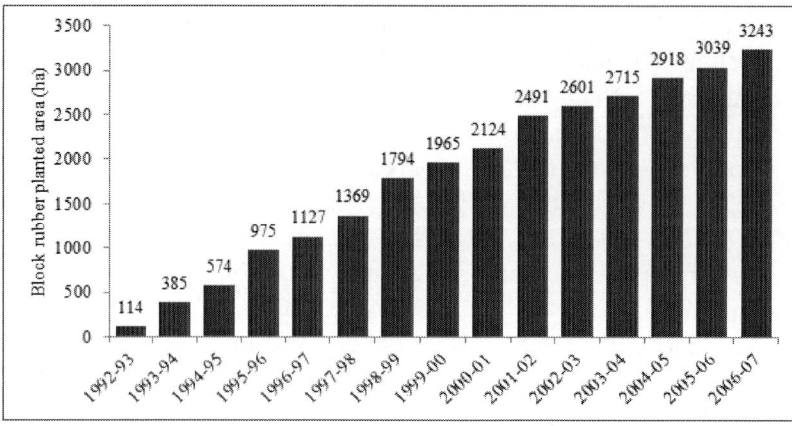

Source: Rubber Board Zonal Office, Guwahati.

Figure 4. Rubber planted Area under block planting scheme in Tripura, 1992-93 to 2006-07.

The average household income from productive (mature) block planting units (BPUs) was Rs. 99168 (US$ 1480) compared to INR (Rs.) 46824 (US$ 699) of immature (young rubber planted) BPUs, which indicates that there was an addition of Rs. 52344 (US$ 781) from rubber after plantations became productive. The annual average value of household savings and assets was Rs. 41966 (US$ 626) and Rs. 183278 (US$ 2735) respectively for members of BPUs after rubber plantations start yielding (Joseph, et al., 2010). The economic life of a productive rubber plantation spans for 18-23 years under normal conditions, unless disturbed by adverse climatic or poor management conditions.

Growth of Monoculture Rubber Plantations in Tripura

Rubber plantations in Tripura are certainly younger compared to those in the traditional regions (Kerala and others) even though it appears to be one of the earliest among the non-traditional areas. Tripura reported the existence of 103 hectares of rubber plantation in as early as 1973. As noted, the initial years witnessed growth due to state initiatives, and response from private sector was slow. However, the institutional support and extension activities of the Rubber Board gave a major boost to the private growers. In fact, the rubber stakeholders in Tripura are from divergent sections and comprise of government agencies, petty business people, agriculturists, traders and government servants. Rubber farms are also owned by residents/firms from other states or regions. Farms are mostly small holdings, though the estates sector plantations account for 20 percent of the total rubber planted area.

It may be seen that within a span of 40 years, the area under rubber monoculture had witnessed a quantum jump to around 60000 ha by 2012-13. The steady increase in the area under monoculture rubber plantations in Tripura is visible in Figure 5. The growth process, however, had witnessed occasional spurts as we see that the volume of the incremental area has not been uniform across the period. The first phase of growth in plantations started in the late 1970s and peaked up in the late 1980s. It was a period of state initiatives in the form of rehabilitation of landless shifting cultivators. The process, as we know, was pioneered by TFDPC and then followed up by TRPC. The growth in the 1990s had a significant contribution from the Block Plantation Scheme of the Rubber Board, which was funded as part of a World Bank Aided Project. However, the boom in the NR prices in the world market has been the primary motivator behind the huge increase in the new plantations in recent years. Almost 14000 hectares of new rubber plantations were raised in the three years beginning 2006-07 and the majority of them were by the private sector.

Table 9 presents the trends in rubber tapped area, production and productivity over the past decade. Tapped area had increased three fold from 11000 ha (2000-01) to 36300 ha (2012-13). Production of rubber increased by more than 3.5 times during the same period from 9980 tonnes to 36300 tonnes in the immediate past year. The production of rubber in Tripura continued growing over the years owing to the fact that more plantations were attaining the stage of productivity. However, the productivity rates exhibited wide fluctuations during the period. The productivity was at a peak in 2004-05, 1165 kgs/hectare. Though there was notable increase in rubber tapped area and production since then, the yield rates have remained lower, recording the next highest at 1089 kg/ha during 2012-13.

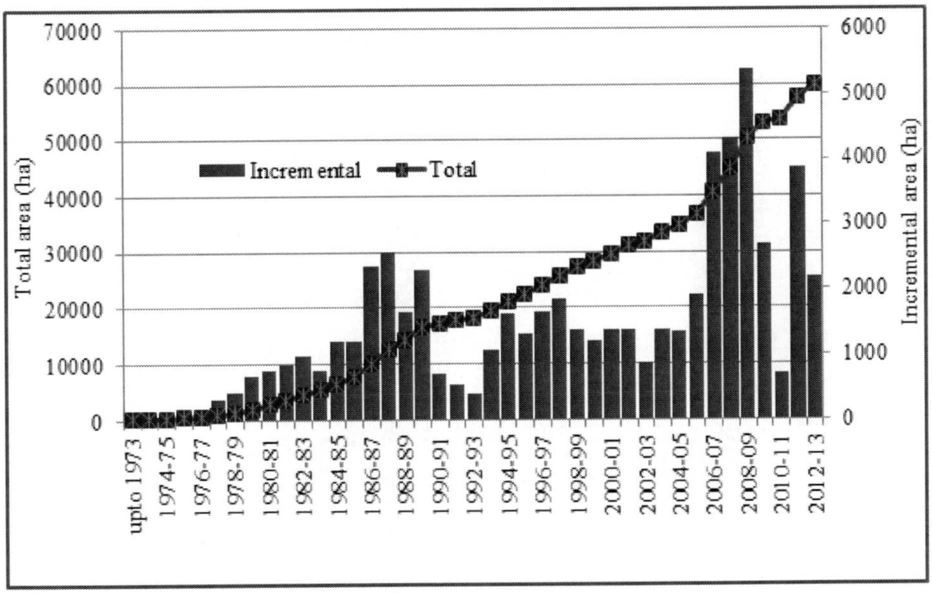

Source: Various issues of Indian Rubber Statistics

Figure 5. Area under Monoculture Rubber Plantations in Tripura (in ha.).

Table 9. Tapped Area, and Production and Productivity of monoculture Rubber in Tripura

Year	Tapped Area (ha)	Production (MT)	Productivity (kg/ha)
2000-01	11000	9980	907
2004-05	13184	15364	1165
2009-10	29737	26810	902
2010-11	31102	29698	955
2011-12	31977	32332	1011
2012-13 (P)	33344	36300	1089

Note: P – Provisional figures.

Source: Rubber Board, Indian Rubber Statistics (various issues); Basic Statistics of Tripura & Economic Review of Tripura.

Table 10. Area under major crops in Tripura between 2004-05 and 2012-13 (hectares)

Crops	2004-05	% share	2012-13	(%) share
1. Rice	256078	52.05	254743	53.70
2. Rubber	34686	7.05	61231	12.91
3. Coconut, Arecanut & Cashewnut	13480	2.74	17776	3.75
4. Pineapple	4980	1.01	11840	2.50
5. Banana	5374	1.09	13580	2.86
6. Orange	2698	0.55	5280	1.11
7. Jackfruit	9032	1.84	9020	1.90
8. Pulses	8071	1.64	8439	1.78
9. Oilseeds	3939	0.80	4814	1.01
10. Tea	8000	1.63	7500	1.58
11. Potato	5280	1.07	8321	1.75
12. Other crops*	140382	28.53	71834	15.14
Total	492000	100.00	474378	100.00

Note: Other crops* - include maize, wheat, cotton, mango, oilseeds, mesta, jute, spices, lemon, fruits and vegetable crops.

Source: Various Issues of Economic Review of Tripura

Status of Rubber in the Agrarian and Socio-Economic Landscape of Tripura

As per the land use statistics of 2012-13, Tripura reported a gross cropped area ratio of 45% out of the total geographical area of 1.05 million ha. The reported forest cover was 60%, recording a 4% increase over 2004-05. Table 10 presents the latest reported cropping pattern in Tripura. Rice is the principal crop in the state while a variety of other crops are also cultivated including fruits and vegetables. Plantation crops like cashew, arecanut and coconut also have long traditions in the state.

Quite interestingly, compared to several other crops, the rise in area under monoculture rubber has been quite significant during the eight year period, as evident from the Table 10. In relative terms, the share of area under monoculture rubber in the gross cropped area had increased from 7% during 2004-05 to almost 13% during 2012-13. The almost doubling of the area under rubber monoculture during the period seems to have corresponded with the decline in area under other crops, including food crops, like wheat along with mesta, cotton and jute, which are mainly grown in the plains lands. While this needs further confirmation, consultations with stakeholders also indicate the felling of various trees and groves, particularly in the highlands, known as 'tilla'(in local parlance), clearing of natural vegetations, age old trees, medicinal plants, etc for growing rubber. There have been several instances of plantation of rubber trees encroaching forest lands, authenticated by numerous reports of forest authorities taking punitive action by cutting down the rubber trees. However, it should be noted that rubber plantations are considered as a non-forestry activity implying that it cannot be grown over Reserve Forest areas. The plantations developed by the state forest department and TFDPC are known to be on degraded forest lands.

Apparently, the adoption of monoculture rubber in the NER also had been reasoned by the economic returns that rubber provided in comparison to the alternative crops and pre-existed land use and livelihood systems in the region. By virtue of the default plantation development programme and the subsidy policy being promoted by the Rubber Board, the rubber growers (irrespective of the size of land put in for rubber) are not supposed to grow any other intercrops inside the rubber plantations except the seven years of immaturity period. In view of this, many crops and land use practices have ceased to exist after they started growing rubber, including the jhum cultivation, which had remained as a habitual practice for the growers for generations together and continued to offer food security benefits. The farmers opted for rubber monoculture as it provides them income all through the year and turned remunerative than many other crops. Reportedly, the annual income accrued from one hectare of rubber monoculture was about INR 90,000-100,000 (US$ 1343-1493 per ha) against INR 45,000 to 50,000 in case of a single species of bamboo plantation as well as INR 10,000 – 25,000 in case of other agricultural crops, including rice, vegetables, etc.[6]. Naturally, these instances reiterate the gaining popularity of rubber monoculture in Tripura. The positive socio-economic externalities of rubber cultivation were also greatly appreciated by the households as evident from a survey undertaken in Tripura. The assessment of the status of the households in the pre and post rubber scenario has been quite revealing in many respects (Table 11).

Table 11. Status of households in Tripura before and after rubber cultivation

Status	Prior to rubber scenario	Post rubber scenario
1. Household Income	INR 500-1500 per month	INR 10000 -15000 per month
2. Main occupation	Agricultural labour, daily labour, cutting forests, selling fire wood, Jhum cultivation	Regular work in rubber plantations, tapping, processing, managing RPS.
3. Attitude and habits	Vulnerable; often misguided by anti – socials.	Focus on development/ socio-economic upliftment
4. Education	Poor, large scale illiteracy	Appreciates the value of education. Improved Literacy
5. Well being	No saving habit, poor health, poor housing, poor clothing, under- nutrition	Savings in bank, buying new lands, cultivation of rubber, better housing, clothing, better nutrition, consumers durables. Owned TVs, Cars and motor bikes.
6. Village Infrastructure facilities	Non- existent	Much improved- brick roads, electricity, drinking water, Angan Wadi Centres and Schools are now available

Note: 1 US$ = 67 INR (2016).
Source: Primary Survey covering 39 rubber growers, 2013.

[6]The profitability of rubber cultivation in the NE region in 2005 was reported as about INR 44000-55000 per ha per annum as against INR 26000 per ha from the existing integrated farming systems comprising livestock, shifting cultivation, rice cultivation and horticulture (Viswanathan and Shivakoti, 2006).

Rubber Monoculture: Health and Environmental Issues

At the same time, the farmer responses also highlight some of the negative externalities they face owing to the raising of rubber plantations in their vicinity (Figure 6). Almost 54% of the 39 respondents apprehend that rubber plantations cause changes in the local weather conditions. Further, one-third of the respondents consider rubber as a cause for increased disease and health problems mainly arising from the foul smell during rubber processing. Most of the negative impacts indicated by the households are related to environment/ecology. In this context, it may be noted that indiscriminate way of growing rubber may affect the micro-climate and deter bio-diversity[7]. Though presence of more than 40 plant species belonging to about 30 different families have been recorded in a rubber plantation (Jacob, 2000), monoculture rubber always runs the risk of diseases and damages, which has not received adequate attention in the policy circles especially in this region, which is quite known for its rich agri-biodiversity. This also calls for more empirical assessments to better understand the interface between rubber plantations and the agri-biodiversity (adverse effects on the species diversity) of the region in particular. It has been reported by the experienced farmers that rubber cultivation has drastically reduced the biological and plant diversity in the plots concerned, apart from crop diversity that existed in the villages before rubber was introduced.

Institutional Interventions for Rubber Development in Meghalaya and Assam

The case of rubber development in Meghalaya has been quite distinct in terms of a rubber grower co-operative, called the Mendipathar Multi-purpose Co-operative Society (MMCS) acting as an intermediary agency in extending the rubber development programmes in the Garo Hills as promoted by the Indian Rubber Board. As may be seen from Figure 7, there was more than three-fold increase in monoculture rubber planted area in Meghalaya during the last decade from mere 4000 ha (2000-01) to about 13,000 ha during 2012-13. The corresponding increase in rubber production was little more than four-fold, which increased from 1700 tonnes to 7100 tonnes during the same period.

The introduction of rubber in the Garo Hills in the late 1980s had brought in drastic transformation in the villages. The Garo Hills districts together occupy more than 70% of the total rubber planted area in Meghalaya with an average holding size ranging from 0.56 ha in East and North Garo Hills to 0.52 ha in West Garo Hills and 0.46 ha in South Garo Hills. The initial responses towards adopting rubber was not very much encouraging among the tribal communities due to lack of awareness about the crop. However, the successful outcomes of regular rubber output and higher profitability of rubber as achieved by the non-tribal rubber growers in the neighbourhoods have motivated the tribal communities to grow rubber (Viswanathan, 2008).

---

[7]It should be noted that there is a general feeling among a large segment of the population in Tripura that the increasing trend of high temperatures in the state is a cause of indiscriminate way of growing rubber. However, these people, inclusive of all sections of the society do not have any proof to substantiate their argument.

The results of the economic performance of the MMCS as well as the income from rubber earned by the tribal rubber growers who are members of Society are shown in Table 12.

It reveals that there was almost three fold increase in the rubber traded by the Society and more than 90% of the rubber traded was in graded sheets with the percentage of scrap rubber becoming almost negligible over time. Currently, the Society trades almost 20000 tonnes of rubber on a monthly basis with 275 rubber growers attached to it. In turn, the average earnings of member growers from rubber cultivation had significantly increased in recent years to the tune of INR 10,000 to 14,000 per month.

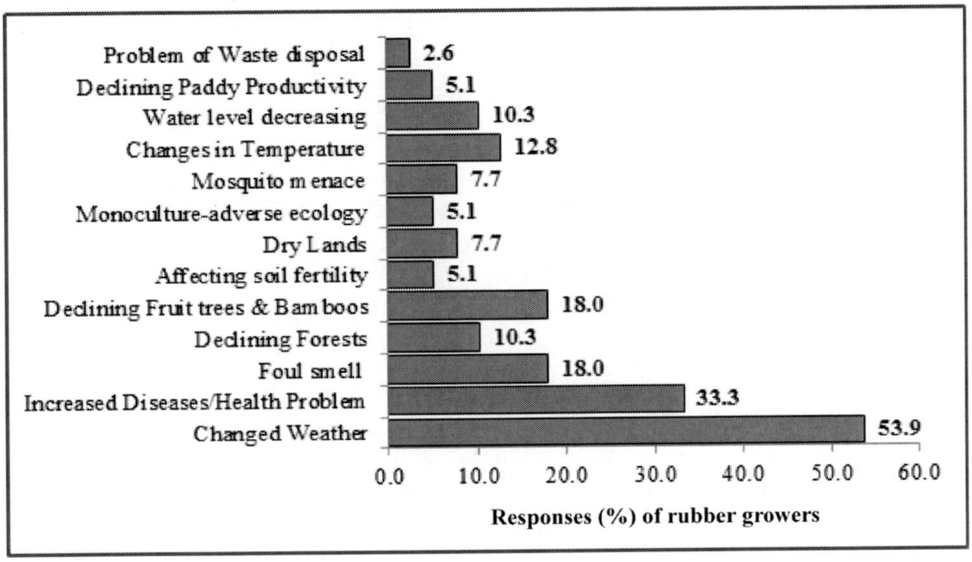

Source: Primary survey covering 39 rubber growers, 2013.

Figure 6. Negative impact of monoculture rubber plantation in Tripura [N = 39].

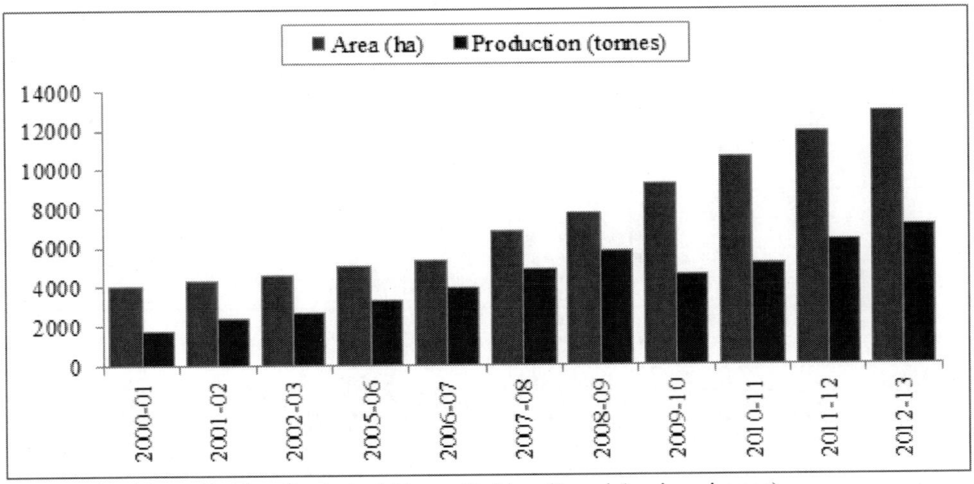

Source: Compiled from Rubber Statistical News, Rubber Board (various issues).

Figure 7. Trends in area and production of Natural Rubber in Meghalaya, 2000-2012.

Table 12. Economic performance of the MMCS and the income from rubber sales received by the member growers [three year averages]

Period	Rubber collected from growers (Tonnes)			Scrap (%)	Rubber traded (monthly) [Tonnes]	Income earned by member growers (INR)	
	Sheet	Scrap	Total			Annual	Monthly
1999-01	70.74	10.61	81.35	13.40	6.78	22022	1835
2001-03	157.60	19.84	177.44	11.34	14.79	38375	3198
2003-05	204.75	21.95	226.70	9.74	18.89	60303	5025
2005-07	174.60	29.08	203.69	14.43	16.97	67042	5587
2007-09	148.70	22.63	171.33	13.07	14.28	66139	5512
2009-11	176.47	14.60	191.07	7.61	15.92	114929	9578
2011-13	216.27	18.77	235.04	8.02	19.59	159844	13320

Source: Mendipathar Multipurpose Co-operative Society, Mendipathar, Meghalaya

It is important to note that the MMCS acts as a catalyst between the rubber growers and the Rubber Board by assuming mediating roles of delivering the planting subsidies as well as supplying farm inputs required for rubber production. This enables the growers to get easy access to the institutional and extension support provided by the Rubber Board through its Regional Office located far away in Guwahati in Assam (Viswanathan, 2008).

Thus, it emerges from the above analysis that adoption of monoculture rubber was certainly beneficial to the tribal communities in terms of providing regular flow of income which was unheard of in the region before the introduction of rubber. It appears that the specific socio-economic conditions of the tribals as well as the market and rural infrastructure facilities prevailed in the Garo villages have been such that they needed immediate attention for improving their socio-economic conditions, which had resulted in the wider acceptance of the rubber development schemes proposed by the Rubber Board.

Table 13. Physical, financial and socio-economic dimensions of rubber development in Assam

Year	Total rubber area (Ha)	Production (MT)	Tappedarea (%)	Employment generated		Productivity (Kg/ha)	Gross value of rubber per ha (Rs.)
				Mandays ('000)	Per ha (persons)		
2006-07	16508	3050	17.65	1504	91	1047	101750
2007-08	18271	4750	18.17	1763	96	1431	139038
2008-09	20975	5097	16.77	2041	97	1449	140798
2009-10	23075	9832	28.19	2710	117	1511	146883
2010-11	27083	10213	28.38	2767	102	1329	129131
2011-12	32117	10950	26.80	3035	94	1272	185723

Source: Authors' estimates based on data gathered from Rubber Board Zonal Office, Guwahati; Employment figures gathered from Economic Survey Assam- 2012-13, Government of Assam.

Table 14. Area, production and productivity of rubber in major districts of Assam, 2010-11

No	District	Total rubber area (ha)	Tapped area (%)	Production (in MT)	Productivity (kg/ ha)
1	Goalpara	7394.08	14.28	1675	1586
2	Karbi Anglong	4174.00	51.02	3386	1590
3	Karimganj	3600.64	35.25	1396	1100
4	Bongaigaon	2045.87	12.33	347	1376
5	Kamrup	1905.00	22.83	645	1483
6	Cachar	1206.69	54.40	884	1347
7	Kokrajhar	1166.57	10.84	155	1226
8	Hailakandi	942.00	54.76	732	1419
9	Dhubri	896.00	17.39	210	1348
10	Chirang	594.91	4.61	37	1348
11	Nagaon	510.88	18.61	156	1640
12	Golaghat	409.29	17.10	105	1500
13	Morigaon	316.00	8.54	40	1481
14	Sonitpur	305.63	35.56	172	1583
15	Darrang	306.00	6.54	30	1500
16	Other 11 districts	1309.01	56.65	243	328
	Total	27082.57	28.38	10213	1329

Source: Economic Survey, Assam 2011; Statistical Handbook of Assam 2011.

The case of Assam also shows significant expansion of rubber within a shorter span of one and a half decades since early 2000s. As of now, Assam holds the second position in rubber area and production in the entire NER with a relative contribution of 31% in total rubber area and 21% in rubber production during 2012-13. As per the latest statistics of the Rubber Board, rubber planted area in Assam has crossed more than 43,000 ha during 2012-13. On an average, the state produces about 6500-7000 tonnes of rubber per annum (Table 13).

The tapped area has seen an increase in recent years, suggesting that more and more rubber plantations are attaining maturity. On an average, the employment generated through rubber plantation development in the state has been about 2.7 million mandays during the latest years reported.

From Table 14, it may be seen that almost all of the districts of Assam have started adopting monoculture rubber, though the extent of area under plantations are meager in size. While Goalpara district has the highest proportion of area under rubber, the proportions of tapped area and production are the highest in case of Karbi Anglong district. In terms of productivity of rubber, Nagaon reports the highest (1640 kg/ha), followed by Karbi Anglong (1590 kg/ha) and Goalpara (1586 kg/ha) districts.

The economic assessment based on discounted cash flow analysis of rubber cultivation in Goalpara district of Assam reveals that the benefit cost ratio had increased from 1.6 during 2005 (Viswanathan, 2006) to 3.1 during 2009-10. The corresponding internal rate of return (at market rate of interest) had increased from 25% to 35%. However, as rubber prices had declined since, there was a decline in the internal rate of return during the past two years, which hovered around 23-24% (Nath and Bezbaruah, 2010; Dharmendra Nath - personal communication dt. 18 March 2014).

Besides the potential threat of rubber monoculture, one of the other major issues facing the rubber production sector in Assam, as revealed from the interactions with farmers is that the farmers with relatively lower size of rubber areas or smaller number of rubber trees enter into contractual arrangements with relatively larger rubber growers for tapping the rubber trees and the larger growers in turn recover about 20-25% of the value of the rubber sold in the market. In most cases, such contracts are in monetary terms. It also reflects on the ineffective institutional interventions by the Rubber Board in the marketing of rubber in Assam.

Compatibility of Policies and Institutional Interventions in Kerala and NER: Issues and Challenges

Thus, the three cases of the interventions for rubber expansion and the development outcomes in the NER bring out the interesting institutional dynamics, raising several concerns about the replicability of the institutional model of monoculture rubber development devised by the Rubber Board suiting to the context of the traditional regions, especially, Kerala. The analysis reveals that rubber development programmes and interventions by the Rubber Board have been mainly focused on increasing area and production in the NER and thereby ensure the supply of rubber catering the ever growing industrial requirements for the raw material. At the same time, it is also evident that the income benefits from the wider adoption of monoculture rubber cultivation by farmers in southern India as well as the tribal communities in the NER have been quite significant and enabled a vast majority to overcome poverty along with notable social and human development outcomes.

However, the question of replicability and compatibility of the policy and institutional interventions for monoculture rubber development as evolved in the traditional regions of Kerala and other south Indian states towards the NER is a point of contention, in view of the potential threats that such a monoculture will cause to the agro-ecosystems and biodiversity of the regions. It may be observed that much of the extant policy and institutional interventions have been evolved over a longer span of 4-5 decades and have been unilaterally focused on promoting rubber monoculture across regions in the country. The institutional interventions and support regime evolved for promoting rubber cultivation in India (and Kerala) has been quite pervasive in terms of its focus on monoculture, as it was thought (with some scientific reasoning) that rubber grown in a mixed cropping system may not yield its fullest potential. Essentially, since many of these policies and interventions have been co-evolved along with the rise and growth of the Rubber Board to facilitate the development and expansion of rubber in the traditional regions, especially Kerala with totally different socio-economic and agro-ecological conditions, the replication of the monoculture model in the NER may have serious consequences in the long run.

In this respect, we present some of the emerging issues and challenges that raise serious concerns about the compatibility of the existing policy framework and institutional interventions for rubber development in the context of the NER. These issues logically arise from the fact that the existing interventions for rubber development have been evolved for facilitating monoculture rubber expansion in the specific context of the traditional rubber growing regions of southern India, especially, Kerala. Hence, there are problems in replicating such development strategies in the NE context. In what follows, we present and discuss some of these issues and challenges:

1. Given the specific socio-economic, ethnic and institutional settings of the NE states, which are also quite diverse in terms of integrated farm livelihood systems combined with shifting cultivation practices (Viswanathan, 2006), it is important that the institutional interventions for monoculture rubber development in the region should be critically examined in view of the sensitivity and agro-ecological diversity of the region and the grower characteristics.
2. The institutional interventions as promoted in the context of the NER seem to have incorporated the specific R&D requirements for monoculture rubber development in the region ignoring the prevailing integrated farm livelihood (crop/ fishery/ livestock) systems as well as the socio-economic, cultural, ethnic, ecological milieu of the communities in the region.
3. Rubber smallholder development in the context of India has been promoted by the national government policies as a monoculture system, which remained completely detached with the possibilities of integration of multiple cropping/ land use systems between rubber and other livelihood systems, as exist in the newly emerging rubber growing areas of North-eastern states, especially, Tripura and Meghalaya. Being one of the most important remaining ecosystems and biodiversity hotspots, the North Eastern regions of the country would require a very distinct model of smallholder rubber development as integrated farm-livelihood systems rather than as a monoculture system as being currently promoted in the region.
4. It is also important to consider that NER is diverse in terms of agricultural as well as livelihood practices, including inland fisheries, upland rice, traditional vegetables, etc. In this regard, development of rubber plantations as a monoculture system may come in conflict with the pre-existing as well as co-existing agricultural production (including food crops) practices/ farm integrated livelihood systems.
5. Reportedly, rubber development in the NER also confronts problems of growing conflicts between the Forest Department and the Rubber Board, as rubber monoculture is also being expanded towards the fringes of forest areas, including degraded forest lands. Many instances are reported in which the Forest Department is not being consulted by the Rubber Board about the rubber development activities and the encroachment into the forest as well as village common lands.
6. From the stand point of economic sustainability of rubber production, a major challenge in the NE states is the growing demand for skilled tappers. It may be noted that on an average, 6000-8000 ha of rubber plantations will become tappable annually. This necessitates that minimum of 2000-2500 new rubber tappers will have to be identified and trained for the task. In the given scenario, there are serious limitations in achieving these targets due to the lack of facilities for giving training to such a number of rubber tappers in a shorter period of time.
7. It is important to explore, identify and implement alternative farming models for commercial rubber cultivation in the northeast that secure the livelihood needs and other aspirations of the ethnic communities, who mainly depend on the land with its indigenous biodiversity, compared to those in the traditional rubber regions in Kerala. Hence, there is a dire need for dedicated trials in the NER to promote rubber

in pure blocks of one acre or more separated by blocks of existing natural vegetation or new plantations of trees bearing edibles and trees for other utilities chosen by the local communities. The farmers should be encouraged for intercropping under immature rubber plantations with tuber crops and legumes to ensure food security and income during the long gestation period by linking subsidy to comply with this planting model. Several typologies of the rubber integrated farming systems being widely adopted by farmers in Thailand and Indonesia may be important to be considered as viable options for the NER with suitable adaptations.

8. Last, but not least, institutional compatibility in the NER context is also important to be understood with respect to the interface between Rubber Board interventions and the interventions by other crop promotional agencies/ institutions in the local settings of the rubber growing areas in the NE states. Though rubber has immense potential for enhancing and ensuring inclusive growth for the tribal communities in the NER, there is a strong case for devising a sustainable land use plan for the region integrating rubber with several combinations of horticulture and other crops along with the existing farm livelihood pursuits (like inland fisheries, livestock, piggery, etc) of the communities. To a greater extent, this would be possible only through networking and collaborative efforts between the various line departments (like Agriculture, Forestry, Horticulture, Rural Development, etc.) as well as the national and regional agencies, such as the ICAR, Spices Board, Coffee Board, North Eastern Council, National Bamboo Mission, State Biodiversity Boards, National Medicinal Plants Board, Coconut Development Board, etc.

CONCLUSION

The policies and institutional interventions promoting rubber development in India have been greatly influenced by the colonial model of monoculture plantation development strategies that still exist in many countries in the South and SE Asian region. Of late, there has been a surge in empirical literature from all across these countries, especially, China, India, Cambodia, Myanmar and Lao PDR, that several decades of intensive promotion of monoculture cultivation of rubber by the governmental as well as other development agencies, have posed significant threats to the agrarian land use systems, agro-ecologies, biodiversity and the rainforest conservation practices (Wu, *et al.*, 2001; Liu *et. al*, 2006; Mahanty, *et al.*, 2006; Majumdar, *et al.*, 2014; Viswanathan, 2014; Chattopadhyay, 2015; Neth Top and Vuthy Lic, 2015). For instance, In Lao PDR, the villagers wanted to plant more rubber since the market began to boom in 2004 but have been limited by scarcity of land for rubber expansion. As a consequence, many forests have been cleared to plant rubber. In Ban Lomue, for example, villagers encroached on the village's protected area to grow rubber. In Ban Oudomsin and Ban Namdeth Mai, villagers are involved in a dispute concerning a conservation forest that was cleared for rubber (Mahanty, *et al.*, 2006). Similarly, in Cambodia, it has been reported that people's livelihoods have changed rapidly after the establishment of rubber plantation in Tumring commune. Crop lands were lost and rice yield in the area decreased after the establishment of the rubber plantation. The restrictions not only limited people's movement, but livestock belonging to the local population was banned from grassland located within the plantation area, with the imposition of a large penalty (Neth Top and Vuthy Lic, 2015).

Interestingly, despite its existence for over six decades, the case of rubber monoculture as well as its long-term implications on the environment and agro-ecologies in the traditional regions of Southern India, especially, Kerala are yet to receive adequate attention in the policy circles. While more scientific empirical studies are yet to address these issues, available studies (Chattopadhyay, 2015; Viswanathan, 2014; Viswanathan, 2015) mainly Kerala) indicate that mono cropping of rubber has already replaced natural vegetation as well as important commercial and other crops in the state, such as coconut, arecanut, cashew, tapioca, banana, and tea. Rubber also had resulted in permanent conversion of significant amount of low lying paddy fields as well as wetlands. There are also evidences of depletion of groundwater table in rubber plantation areas (Shaji, 2011; Karunakaran, 2014) as well as contamination of streams, rivers and other water bodies as caused by the flow of effluents from rubber sheet processing activities (Sreelakshmi, *et al.*, 2007) in the rubber dominated areas. It is also reported that due to complete coverage, penetration of sunlight is restricted under canopy rubber plantation, adversely affecting the growth of other crops in and around the rubber growing areas. Apart from these, there are also problems that are normally associated with monocropping, like drawing on same nutrients with uniform rooting system and susceptibility to various plant and leaf diseases in rubber (Chattopadhyay, 2015). However, these issues still remain grossly under-reported in the state and, hence, there is an urgent need for detailed investigations across rubber growing regions. There are also concerns emerging from rubber area far exceeding the agro-ecological limits imposed by the geographical features of the Kerala state. For instance, it has been observed that rubber monoculture area has already grown out of proportion in relation to the proportion of lowland and wetland areas across districts in the state (Viswanathan, 2014).

Besides these environmental challenges posed by the development of monoculture rubber, the recent decades also witnessed tremendous growth in investments in rubber plantations across these countries. In many cases, these investments involve massive land grabbing[8] by large private and corporate enterprises for establishment of monoculture plantations. There are also several instances of cross border investments in rubber monoculture by Chinese as well as Vietnamese companies in Lao PDR, Cambodia and Myanmar (Byerlee *et al.*, 2014). The environmental as well as agro-ecological consequences of development of such monoculture rubber plantations would assume alarming proportions in the future unless appropriate policies and sustainable rubber development strategies are devised at the regional level to contain the onslaught of these environmentally destructive investments.

Given these environmental challenges and threats posed by the rubber monoculture in the larger context of the Asian region in general as well as in the regional contexts of Southern and North Eastern states in India, it is important that the monoculture oriented rubber expansion policies and institutional interventions need a critical examination and recasting to make rubber cultivation a sustainable agriculture system along with its integration with other cropping practices as well as livelihood activities. In this context, there is an urgent need for a paradigm shift in policies and interventions for rubber development in the regional context of

[8]In South East Asia, large-scale rubber plantations are one of the main drivers of 'land grabs' and deforestation. Land deals don't need to happen this way – there is significant evidence to show that investing and supporting smallholder farmers brings lasting economic, social and environmental benefits (Global Witness, 2014).

S& SE Asia, towards rubber integrated farm business and livelihood systems (RFBLS). A case in point here is the *rubber agro-forestry systems*, also known as "*jungle rubber*" as exist in Indonesia[9]. There has been ongoing research and efforts to promote rubber agro-forestry systems (RAS) in Indonesia and Thailand in particular, due to the socioeconomic and environmental benefits offered by this approach. These benefits include: (a) reduced risks for smallholders emanating from persistent decline and volatility in rubber prices; (b) increased biodiversity; and (c) environmental and ecosystem benefits as emerging from integrated farming systems. In the fast changing global economic as well as trading environments along with ensuing threats from climate change and increasing price risks, sustainable rubber promotion strategies should also give adequate attention towards setting newer production standards that are environmentally friendly as well as ecologically compatible. As indicated by the most recent developments, there is already a growing demand from both the consumers and the producers for green certified rubber, in view of the growing consumer awareness of the loss of biodiversity through monoculture farming.

ACKNOWLEDGMENTS

The authors thank the National Research Programme on Plantation Development (NRPPD) at the Centre for Development Studies, Trivandrum, India for supporting this study. We also thank the Senior Officials of the Directorate of Agriculture, Government of Assam; Government of Tripura and Dr. Kalyan Das for the useful interactions and inputs provided during the course of the study. The usual disclaimers apply.

APPENDIX 1: KEY INDICATORS OF AGRICULTURE SECTOR IN NE STATES, 2010-11

State/ Region	Total cultivated area('000 ha)	Arable land per capita (ha)	Share of agriculture in GSDP (%)	(%) of Agricultural population			Marginal & Small farmers (MSF- %)	Fertlizer use* (kg/ha)	Rice area (%)
				Cultivators	Agri. Lab.	Total			
Arunachal	253	2.93	31.06	51.5	6.2	57.7	37.3	2.83	34.6
Assam	2889	1.24	27.31	33.9	15.4	49.3	85.6	65.98	77.2
Manipur	348	1.64	24.77	39.5	9.6	49.1	83.4	37.82	68.6
Meghalaya	342	4.94	18.06	41.7	16.7	58.4	81.3	14.05	41.8
Mizoram	197	2.87	18.79	47.2	8.6	55.8	87.0	37.39	60.2
Nagaland	417	4.47	24.99	55.2	6.5	61.7	15.8	2.74	55.8

[9] Jungle rubber refers to rubber agroforests, or "*jungle rubber*", as prevail in Indonesia, Malaysia and Thailand since the first half of the 20th century. Farmers incorporated rubber in their traditional "slash-and-burn" practices, allowing rubber to grow with natural secondary vegetation, as well as timber and fruit trees. Although most complex rubber agroforests have disappeared in Malaysia and Thailand, and are under threat in Indonesia, there were still more than 1 million ha of jungle rubber in Indonesia at the beginning of the 1990s (Gouyon, et al., 1993).

APPENDIX 1: (CONTINUED)

State/ Region	Total cultivated area('000 ha)	Arable land per capita (ha)	Share of agriculture in GSDP (%)	(%) of Agricultural population			Marginal & Small farmers (MSF- %)	Fertlizer use* (kg/ha)	Rice area (%)
				Cultivators	Agri. Lab.	Total			
Sikkim	82	3.02	10.84	38.1	8.4	46.5	76.6	NA	11.3
Tripura	258	0.65	19.25	20.1	24.1	44.2	94.9	53.51	56.3
NER	4786	2.72	21.88	40.90	11.94	52.84	70.2	51.39	50.7

Note: GSDP – Gross State Domestic Product at 2010-11 prices. * - Fertilizer use relates the averages for the period 2009-10 to 2011-12. Arable land per capita relates to 2005-06. Rice area is percentage of total cultivated area during 2010-11. MSF – Marginal and Small Farmers.

Source: GOI, Agricultural Statistics at a Glance 2012; State of Indian Agriculture, 2012-13.

REFERENCES

Bahuguna, Dr. V. K. (2005). Action Plan for Expansion of Rubber in Tripura,submitted to the Government of Tripura.

Bhowmik, I. (2006). “A Status Report on Rubber Plantations in Tripura”, in V K Bahuguna (ed) Natural Rubber in Tripura, Baseline Data & Status Report: Agartala, Government of Tripura, pp. 59-81.

Bhowmik, I. (2008). Unpublished Ph.D thesis- Problems and Prospects of Natural Rubber Production in India, Visva Bharati University.

Bhowmik, I. (2009). “An Insight to the Rubber Economy of Tripura” in KN Jena, et al. (ed) Development Paradigm and Bottom up Approaches: New Delhi, Abhijit, pp. 31-51.

Barlow, Colin. (1997). ‘Growth, structural change and plantation tree crops: The case of rubber’, *World Development*,*25*(10), 1589-1607.

Byerlee, Derek. (2014). ‘The Fall and Rise Again of Plantations in Tropical Asia: History Repeated?’, *Land* (Forthcoming).

Byerlee, Derek., Dolly, Kyaw., U San, Thein.& L Seng, Kham. (2014). ‘Agribusiness Models for Inclusive Growth in Myanmar: Diagnosis and Ways Forward’, Michigan State University (MSU), Myanmar Development Research Institute-Center for Economic and Social Development (MDRI-CESD) &United States Agency for International Development (USDA), *First Draft Report*, 15 February 2014, 76p.

Chakraborty, S. (2012). “Rubber transforms life of poor tribals in Tripura”, sujit.c@ians.inhttp://www.readandknow.org/rubber-transforms-life-of-poor-tribals-in-tripura, IANS 28th October 2012.

Chattopadhyay, Srikumar. (2015). ‘Environmental Consequences of Rubber Plantations in Kerala’ NRPPD Discussion Paper No 44, National Research Programme on Plantation Development, Centre for Development Studies, Trivandrum, Kerala, 54p. http://cds.edu/wp-content/uploads/2015/07/ 44_Srikumar.pdf.

Directorate of Economics & Statistics, (2006). ‘Economic Review of Tripura 2004-05’, Government of Tripura, Agartala.

Directorate of Economics & Statistics, (2013). 'Economic Review of Tripura 2011-12', Government of Tripura, Agartala.

Directorate of Economics & Statistics, (2014). 'Economic Review of Tripura 2012-13', Government of Tripura, Agartala.

DONER, (2007).North Eastern Region Vision 2020,Development of North Eastern Region, Ministry of Development of North Eastern Region, Government of India, New Delhi.

Fernandes, Walter. (2004). 'Limits of Law and Order Approach to the North-East', *Economic and Political Weekly*, *39* (42), (4609-4611 (October 16-22, 2004).

Ganguly, J. B. (1969). Economic Problems of the Jhumias of Tripura, Bookland, Kolkata.

George, K.T., Haridasan, V.& Sreekumar, B. (1988). Role of government and structural changes in rubber plantation industry in India, *Economic and Political Weekly*, *23*(48),M158-166.

GOK, (2011).Economic Review 2010, Government of Kerala, State Planning Board, Thiruvananthapuram.

Gouyon, A., de Foresta, H.& Levang, P. (1993). Does Jungle Rubber trees deserve its name? An Analysis of rubber agro-forestry, *Agroforestry Systems*, *22*, 181-206.

Guha, Amalendu. (1991). Medieval and Early Colonial Assam: Society, Polity, Economy, Calcutta, K.P. Bagchi (on behalf of the Centre for Studies in Social Sciences).

Hayami, Yujiro. (2010). 'Plantations Agriculture', *Handbooks in Economics.*, *18*, 4, pp. 3305-3322.

Hicks, Charlotte., Saykham,Voladeth., Weiyi, Shi., Zhong, Guifeng., Sun, Lei., Pham Quang, Tu.& Marc, Kalina. (2009). "Rubber investments and market linkages in Lao PDR: approaches for sustainability", The Sustainable Mekong Research Network (SUMERNET), 170p (February).

Karunakaran, N. (2014). Crop Diversification and Environmental Conflicts in Kasaragod District of Kerala, *Agricultural Economics Research Review*, *27* (2), 299-308 (July-December), http:// ageconsearch.umn.edu/ bitstream/196987/2/15-N-Karunakaran.pdf (accessed 12 January 2016).

Mahanty, Sango., Jefferson, Fox., Michael, Nurse., Peter, Stephen.& Leslie, McLees. (2006). Hanging in the Balance: Equity in Community-Based Natural Resource Management in Asia, *RECOFTC and East-West Center*, 2006.

Jacob, James. (2000). "Rubber Tree, Man and Environment" in George P.J. &C.K.Jacob (Eds.) Natural Rubber: Agro Management and Crop Processing, Rubber Research Institute, Kottayam, pp. 599-610.

Joseph, Joby., Tharian, George.& Dey, S. K. (2010). 'Report on the Socio-Economic Impact of Natural Rubber Cultivation under the Block Planting Scheme in Tripura', *Working Paper ER/4*, Rubber Research Institute of India, Kottayam, Kerala, pages, xii+39.

Joshi, L., Wibawa, G., Vincent, G., Boutin, D., Akiefnawati, R., Manurung, G., Noordwijk M. V.& Williams, S. (2002). Jungle Rubber: a traditional agroforestry system under pressure, ICARF-World Agroforestry Centre, Indonesia, 2002, 38p.

Lipton, Michael. (2009).*Land reform in developing countries: property rights and property wrongs*,New York: Routledge.

Liu, W., Hu, H., Ma, Y.& Li, H. (2006).*Environmental and Socioeconomic Impacts of Increasing Rubber Plantation in Menglun Township, Southwest China,Mountain Research and Development*, *26* (3), pp. 245-253.

Majumder, Abhik., Sadrita, Datta.,Bal, Krishan Choudhary.& Koushik, Majumdar. (2014). Do Extensive Rubber Plantation Influences Local Environment? A Case Study From Tripura, Northeast India, *Current World Environment*, *9*(3), 768-779.http:// www.cwejournal.org/ vol9no3/ do-extensive-rubber-plantation-influences-local-environment-a-case-study-from-tripura-northeast-india/.

Mohapatra, A. C. (1994). Locational Pattern of Markets in Pre and Early British Period in North-East India (Khasi, Jaintia and Garo Hills), In: J.B. Bhattacharjee (Ed.): *Studies in the Economic History of North East India*, HarAnand Publications, New Delhi, pp. 160-168.

Nair, M.K. S (1986). Constraints on the Development of Land Market in Meghalaya, *Economic and Political Weekly*, *21* (27), PE-60- PE-67.

Nath, Dharmendra and M.P. Bezbaruah (2010): 'Income Augmentation in Small Farm Agriculture form Adoption of Rubber Production', *YSA*, 2010, 109-132, www.sagw.ch/dms/agrarwirtschaft/.../4.../Nath_Bezbaruah.pdf (accessed 29 March 2014).

Neth, Top., Dararath, Yem. &Vuthy, Lic. (2015). Cost-Benefit Studies of Natural Resource Management in Southeast Asia, in Rubber Plantation Development in Cambodia: A Cost-Benefit Analysis, pp. 137-156, Cost-Benefit Studies of Natural Resource Management in Southeast Asia, Editors: David James, Herminia A. Francisco, ISBN: 978-981-287-392-7 (Print) 978-981-287-393-4 (Online).

Ostrom, Elinor. (1990). Governing the Commons: The Evolution of Institutions for Collective Action, New York, NY: Cambridge University Press.

Ostrom, Elinor. (2005). Understanding Institutional Diversity, Princeton, NJ: Princeton University Press.

Ostrom, Elinor. (2007). "Institutional Rational Choice: An Assessment of the Institutional Analysis and Development Framework", in Sabatier, Paul (ed.), 2007, *Theories of the Policy Process*, Boulder, CO: Westview Press.

Paribalan, P. (2006).Role of Rubber Board in Tripura, in V K Bahuguna (ed) Natural Rubber in Tripura, Baseline Data & Status Report. TRM/ Technical Bulletin 1, Agartala, pp. 41-51.

Pereira, Ignatius. (2009). "Rubber erases unrest in Tripura tribe", http:// www. hindu.com/2009/12/24/stories/2009122454360400.htm.

Planning Commission, (2001). 'Report of the Working Group on Agricultural Development in Eastern & North Eastern India for the Formulation of the Tenth Five Year Plan', TFYP Working Group, Sr. No. 40/2001, pages: xi+117.

Prabhakara, M.S. (2004). 'Is North-East India Landlocked?', *Economic and Political Weekly*, *39* (42), 4606-4608 (October 16-22).

Rubber Board, (2013). Rubber Growers Companion- 2013, Kottayam, Govt. of India.

Shaji, E. (2011). 'Groundwater quality of Kerala – Are we on the brink?', paper presented at the Disaster, Risk and Vulnerablity Conference 2011, School of Environmental Sciences, Mahatma Gandhi University, *India in association with the Applied Geoinformatics for Society and Environment*, Germany, March 12–14, 2011, Kottayam, Kerala, http:// www. indiawaterportal.org/sites/indiawaterportal.org/files/disaster_research_shaji_e.pdf.

Shimray, U. A. (2004). 'Socio-Political Unrest in the Region Called North-East India', *Economic and Political Weekly*, *39* (42), 4367-4643 (October 16-22).

Sinha,A. K. (2007). Prospect of rubber plantation in NE region with special reference to the state of Tripura accessed from www.tripurainfo.com on 27/10/2008.

Somboonsuke, B., Shivakoti, G. P.&Demain, H. (2001). 'Agricultural Sustainability through the Empowerment of Rubber Smallholders in Thailand', *Asia-Pacific Journal of Rural Development, 11* (1), 65–89.

Viswanathan, P.K. (2002). 'Irrigation and Agricultural Development in Kerala: An analysis of missed linkages', *Review of Development and Change*, July December 2002, *7* (2), 279-313.

Viswanathan, P. K. (2006).A Comparative Study of Smallholder Rubber and Rubber integrated Farm Livelihood Systems in India and Thailand, Report of the postdoctoral research study submitted to the Asian Institute of Technology (AIT), Bangkok, January 2006, 108p.

Viswanathan, P.K. & Ganesh, P. Shivakoti. (2007). 'Conceptualising Sustainable Farm-livelihood Systems in the Era of Globalisation: A study of Rubber Integrated Farm Livelihood Systems in North East India', *Social Change and Development*, Volume *5*, pp.111-142 (November).

Viswanathan, P.K. & Ganesh, P. Shivakoti. (2008). Integration of Tribal Economies in the Era of Globalisation: Perspectives on Rubber Based Farming Systems in North East India, *Asian Profile*, *36*(3), 281-298, Journal of Asian Research Service, Canada.

Viswanathan, P.K. (2011). 'Co-operatives and Collective Action: Case of a Rubber Grower Co-operative in East Garo Hills in Meghalaya, North East India', *Social Change and Development*, *8*(1), 89-125 (July).

Viswanathan, P.K. (2012). 'Integrated Rubber Farming and Livelihood Systems in Northeastern India', in: Sumi Krishna (ed.), *Agriculture and Changing Environment in Northeastern India*, New Delhi/UK: Routledge (Taylor and Francis Group), pp. 263-288.

Viswanathan, P. K. (2013).*Lao Census of Agriculture 2010/11: Analysis of Selected Themes*, Unpublished Report Submitted to the Ministry of Agriculture and Forestry, Lao People's Democratic Republic, Government of Lao PDR, Vientiane and the Food and Agriculture Organization of the United Nations (FAORAP), Bangkok, 145p.

Viswanathan, P.K. (2014). 'Rationalisation of Agriculture in Kerala and its Implications on Natural Environment, Agro-Ecosystems and Livelihoods', *Agrarian South: Journal of Political Economy*, *3* (1) [in print].

Viswanathan, P. K. &Amita, Shah. (2012). '*Gender Impact of Trade Reforms in Indian Plantation Sector: An Exploratory Analysis*', Discussion Paper No. 17, Centre for Development Studies, Trivandrum, Kerala, 69p. http://cds.edu/admin/homeFiles/nrppd %2017.pdf.

Viswanathan, P. K.&Amita, Shah. (2013). 'Trade Reforms and Crisis in India's Plantation Industry: an analysis of Tea and Rubber Plantation Sectors', *Social Change and Development*, *10* (2), 31-85 (July).

Viswanathan, P. K. (2015). Farmer Migration and Agrarian Transformation in Kerala: Impacts on the Local Economy, Society and the Agrarian Landscape in the Erstwhile Malabar Region,*NRPPD Discussion Paper 43*,38p., Centre for Development Studies, Trivandrum http://cds.edu/wp-content/uploads/2015/07/43_ViswanathanII.pdf.

Woods, K. (2012). 'The Political Ecology of Rubber Production in Myanmar: An Overview', 66p, https:// www.globalwitness.org/ sites/default/ files/ The%20Political%20Ecology %20of%20Rubber%20Production%20in%20Myanmar%20FINAL.pdf.

Wu, Zhao-Lu., Hong-Mao, Liu.& Lin-Yun, Liu. (2001). Rubber cultivation and sustainable development in Xishuangbanna, China, *International Journal of Sustainable Development & World Ecology*, *8*, 4, 337-345.

In: Monoculture Farming
Editors: Tapan Kumar Nath and Patrick O'Reilly

ISBN: 978-1-63485-166-4

Chapter 2

SOCIO-ECONOMIC AND ENVIRONMENTAL IMPACTS OF TEAK MONOCULTURE PLANTATIONS IN SRI LANKA

Mangala De Zoysa*
Department of Agricultural Economics, Faculty of Agriculture,
University of Ruhuna, Mapalana, Kamburupitiya, Sri Lanka

ABSTRACT

The demand for sawn timber or hardwood is growing rapidly in Sri Lanka with the growth in population, economic development, and external trade of timber industry. Teak has become one of the most valuable hardwoods in Sri Lanka, which is paid premium prices at local and international markets due to its beauty, strength, durability and weather-resistant qualities. Teak growth is rapid and there is a relatively rapid turnover and produce homogeneous economic crops. The benefits of teak are in the form of goods: timber, poles, etc. and services: amelioration of microclimate, watershed, reduction of soil erosion, provision for shelter and shade, etc. Commonly, teak monoculture plantations in Sri Lanka has often been criticized for focusing too much on the management of timber and for insufficient attention on ecological functions. Generally, monoculture forest plantations damage environment including the soil, water, and even the climate on an unprecedented scale. However, the commercial teak timber monoculture plantations have a number of positive as well as negative local environmental, social, and economic impacts. As the demand for teak continues to increase in Sri Lanka it is practically vital to investigate the teak monoculture plantations in terms of socio-economic possibility as well as a sustainable solution for growing environmental issues. Hence, the study attempts to review available literature and analysis teak monoculture plantations in Sri Lanka in terms of: teak monoculture plantation strategies; merits and demerits of monoculture teak plantations: and sustainable solution for growing environmental issues of monoculture teak plantations. The main teak monoculture plantation strategies in Sri Lanka are identified as modern industrial plantations; farmer woodlots programs; and home-garden development programs. The merits of monoculture teak plantations are discussed in terms of:

* Corresponding author: mangalaxyz@yahoo.com; mangala@agri.ruh.ac.lk.

reclamation of degraded lands; select most profitable tree species; greater timber yields under less pressure from other species; uniform tree structure; assure adequate supplies of timber; employment and income generation; foreign exchange savings; conservation of natural forest; carbon sequestration and capacity-building for sharing GHG benefits. The demerits of monoculture teak plantations are analyzed through: fail in protecting biodiversity in forest ecosystems; limitation to produce food with shrinking terrestrial resources; vulnerability of monocultures to disease and insects, and greater instability of production; face risks of fire; little or no forest conservation; greater risk under changing environmental conditions; impact on the environment on an unprecedented scale; negative impact on local communities and their livelihoods; and prediction of excessive supply and price declines. Sustainable solution for growing environmental issues of monoculture teak plantations are suggested as: agro-forestry initiatives; introduction of silvopastoral systems; popularization of home-gardening; and managed biodiversity protection functions.

BACKGROUND

The forest industry is of major ecological, economic and social importance in many regions of the world. The industrial forest plantations are promoting monocultures as forestry at the expense of all other forest functions. Monoculture plantations damage environment including the soil, water, and even the climate on an unprecedented scale. The phenomenon of yield decline with monoculture is commonly discussed in terms of replant problem, autotoxicty, and monoculture injury. Cultivation of Monoculture plantations is considered as not sustainable, because decline of yield is due largely to soil-borne plant pathogens adapted specialized for the roots of the trees.

The situation has worsened when the concept and practice of tree monocultures was introduced to tropical regions where the tropical forest characterize tremendous biodiversity and the complexity of interactions (Pancel, 1993). Although the rational modern forestry is largely based on monocultures, forest in general contributes more ecosystem services such as biological diversity, carbon storage, and non-timber forest products than timber production. Tree monoculture plantations, like other agricultural monocultures also have undergone intensive technical development during the last decades. Teak (*Tectona grandis*) is native to the Asia monsoon forests and predominantly grown in monoculture plantations in the tropical latitudes. In addition to the present area of about 3 million ha of teak, there are ample opportunities for investments in teak and other tropical hardwoods to satisfy current world hardwood demand of 90 million m3/year from about 9 million ha of plantations.

From the total land area of Sri Lanka, 28.8% or about 1,860,000 ha covered by trees. The demand for sawn timber or hardwood is growing rapidly in Sri Lanka with the growth in population, economic development, and external trade of timber industry. The country has more plantations with 185,000 ha of planted monoculture tree plantations than primary forest covering 167,000 ha, the most biodiverse and carbon-dense form of forest. Even-aged monocultures of Teak, Eucalyptus or Pine are mostly consisted in plantation forests in Sri Lanka. Teak is native to the Indian-Burmese floristic region and found naturally in India, Myanmar, Thailand and Lao. Dutch in the 17th century introduced teak to Sri Lanka and the first teak plantation was started in 1680. Again, teak was introduced from India and Burma to Sri Lanka, and became popular and prevalent wood during the British period. The first teak planting by the Forest Department in Sri Lanka is dating back to 1872. Teak reforestation was carried out in Sri Lanka under the Taungya System since 1950's and restricted to shifting cultivation lands in 1970's (Ranatunga, 1972). Teak plantation is exclusively planted for the purpose of forestry management, for either commercial or ecological purposes. The commercial teak timber plantations have a number of environmental, social, and economic impacts. The benefits of teak are identified in the form of goods: timber, poles, etc. and services: amelioration of microclimate, watershed, reduction of soil erosion, provision for shelter and shade, etc. Teak one of the most valuable hardwoods is paid premium prices at international market due to its beauty, strength, durability and weather-resistant qualities (Centeno, 1997). Teak monoculture plantations have often been criticized for focusing too much on the management of timber and for insufficient attention on ecological functions. As the demand for teak continues to increase, teak plantations need to provide both socio-economic possibility as well as a sustainable solution for environmental issues. However, management of monoculture teak forest plantations in Sri Lanka has become a complex exercise involved with socio-economic and ecological aspects. Hence, the study attempts to review available literature and analyze the prevailing social, economic and environmental impacts of monoculture teak plantations in Sri Lanka.

Teak Monoculture Plantation Establishments in Sri Lanka

Small and Medium Scale Teak Monoculture Plantations

Industrial teak plantations which are directly managed by the government and smallholder plantations spread across the country which are managed by farmer participation, are the two major teak monoculture plantations available in Sri Lanka. Although teak has been ranked as a high priority species, planting programs are still limited to a small or medium scale and grown as an exotic species. Following the adoption of National Forest Policy in 1980, a variety of community based forest management projects have been implemented in Sri Lanka with donor support since 1982 and attempted to increase tree plantings and thereby reduce poverty and rehabilitate environmentally degraded areas. The Community Forestry Project (CFP) and Participatory Forestry Project (PFP) launched in 1982 and 1993 respectively were the largest operated through Forest Department (FD) with financial assistance from the Asian Development Bank (ADB) (Skutsch, 1990). The community forestry approaches are mainly focused on afforestation, reforestation, agroforestry, homestead garden planting, miscellaneous planting and biodiversity conservation (De Zoysa and Inoue, 2007).

Under different components of the Participatory Forestry Project (PFPs), Forest Department planted teak trees as well as established teak plantations under the popular method of Taungya System, with the aim of sharing the benefits with local people. Teak is the most popular tree species used in different reforestation programs of the Forest Department in Sri Lanka to offset the denudation of natural forests (Ariyadasa, 2003). The Participatory Forestry Project (PFP) commenced in 1993 was mainly focused on Farmer Woodlots (FWLs) on degraded government land, using agroforestry approach for both promoting a wood supply and improving their livelihoods; a well as Protective Woodlots (PWLs) for soil and water conservation and rehabilitation of erosion-prone government land by local communities (ADB 2003). The project identified the land, selected the beneficiaries, blocked out lands and distribution among beneficiaries. A total 4,243 ha of PWLs with tree species including teak were established by 13,773 farmers during the entire project period (Nanayakkara, 2001). Farmer woodlot development programs are mainly considered as an effective means of reforestation by a mixture of temporary cropping with tree establishment. The FWLs of the PFP were established in the lands being used for shifting cultivation, with a 3-year crop rotation. FWLs models were based on an average 0.4 ha plot planted with predominant wood tree species mainly teak (ABD 2003). Forest Department had established about 15,500 ha of Farmers' Woodlots (FWLs) mainly with teak component in 19 districts of Sri Lanka under the Participatory Forestry Management Project implemented during 1993 ~ 1999. Many FWLs programs were initiated in Sri Lanka also in 2002 where small blocks of state land are given on a 25- year period lease to the local community to plant both forest trees mainly teak and temporary cash crops initial 3 years. Teak cultivations under farmer woodlots programs are becoming popular because of the rising realisation of its potential to contribute to socio-economic development of the farmers and also in resolving conflicts that are arising because of environmental scarcities.

Modern Industrial Plantations

Monoculture tree growing genetically similar identical trees over a large areas and produces greater yield under less pressure from other species and more uniform trees structure which is widely used in industrial teak plantations. Government managed teak forest plantations are dominated by industrial forestry, the system of management intensive wood production, featuring enormous monoculture forest plantations. Further, many other teak plantations have also been established in areas with rather poor tertiary soils and on slopes.Total teak forest plantations under Forest Department in 2012 were 31,678 ha (Ruwanpathirana, 2012).

Private Teak Plantations

Department of Forest Conservation in Sri Lanka manages about 94% of the total forest plantation (Pitigala and Gunatilake 2002). Private sector until recent past did not play any significant role in forest plantation. National Forest Policy in 1995 facilitated the establishment and management of industrial forest plantations on the state lands entrusted to local communities, industries and other private bodies. Policy and legal framework conducive to large scale private sector investments in forest plantation development have been provided by Current National Forest Policy of the country (Ariyadasa, 2003). Government has promoted private sector participation for the establishment of forest plantations in order to increase the tree cover and to save the government funds for other national purposes. Government is making attempts and recognized the necessity of supporting policy, legislation and other support systems for popularizing private sector investment in forest plantations.

Most of the private sector investments in timber and industrial wood production have concentrated on teak monoculture plantations. The private plantation companies who export their timber have more concern about teak Plantations which are mainly managed for commercial purposes. The Sri Lankan government's Board of Investment (BOI) which was established in 1992, approved AsiaTeak in December 2010 as reputable foreign operator company. AsiaTeak has invested in teak plantations in Sri Lanka where the best in the world for growing teak produce excellent grain and weather hardiness in the heartwood, the qualities bulk buyers are looking for. Sadaharitha Plantations Limited also engages in establishment of teak plantations to generate investment opportunities for Sri Lankans and the Sri Lankans who are working or living away from the motherland. They are committed for sustainable forest management while conserving environment, protecting healthiness of community through the company processes, products, and services. Presently, several other private companies also have started Teak, Mahogany and White Sandal wood reforestation programs. Private investors have to use the most intensive management principles to achieve their expected growth of teak plantation around 20 years which is significantly lower than the state plantations (Subasinghe, 2007).

Economic Impacts of Teack Monoculture Plantations

Greater Timber Yields

Compared with the mixed species, monoculture teak plantations in Sri Lanka produce more timber volume, and better and uniform quality timber from the same area of land. Teak is one of the best values with the highest yields in the timber sector in Sri Lanka. Minimum rotation for teak is 20 to 25 years. Growth rates over a plantation with an overall average of 12 to 14 cubic meters per year. Yield after 25 years of growth in a properly managed plantation is about 200 to 220 cubic meters per hectare. Monoculture teak forest plantations have ability to grow well under the specific conditions of a particular place. Teak has been classified as a pioneer species. Hence, it requires a high light intensity for its growth and development (Kadambi, 1972). As a pioneer species teak requires a high light intensity particularly, optimum light intensity required for growth and development is between 75-90% (Kaosa-ard, 1999). Monoculture plantations reduce competition between tree species for soil nutrients, water and sunlight, accelerating the growth of the stem diameter. Age class distribution and growing stock of teak in monoculture industrial plantations belong to the Forest Department in Sri Lanka are given in the Table – 1.

Uniform Tree Structure

The length of the clear bole is the most important form characteristic determining the value of teak logs. The length of the clear bole of teak monoculture plantations in Sri Lanka maximizes the value of the log for high-quality end uses. The stem straightness is strongly controlled by provenance used for monoculture teak plantations and is thus genetically inherited character. The mean heritability value of stem straightness was found to be 0.83 in an international provenance Trial (Kaosa-ard, 1999). Various silviculture research programmes have been implemented in Sri Lanka over a long period of time. Species selection by field trials, particularly exotic species mainly teak, facilitated for establishment of large scale commercial teak plantations. Subsequent provenance testing has further improves the performance of these teak species. Genetic improvement programs are focused on Teak, Eucalypts and Mahogany as the main commercial tree species (http://journals.sjp.ac.lk/index.php/fesympo/article/viewFile/57/29). Tissue culture as the technique of propagation is an important technically and economically feasible option for mass supplying of genetically improved materials for the planting program. Supply of genetically improved materials for large scale planting programs is the major problem in establishment of teak plantations in Sri Lanka.

Further, the intensive management strategies of teak monoculture plantations particularly the spacing of trees and the number, timing and intensity of thinning strongly influence the pattern of growth and maximize the length of the clear bole. Teak trees in the monoculture plantation need carefully thinned, so they grow straight for resulting in a better-quality wood. Periodic thinning is highly recommended by the Forest Department for teak monoculture plantation in order to remove slower growing, diseased or poorly formed trees, and to allow the faster growing, better quality trees to achieve their growth potential. Therefore, initially 1,100 trees are planted per hectare and the stand is periodically thinned for quality and to maximize growth rates. Only around 260 trees per hectare are maintained by the harvest time. Although the Forest Department recommend early and heavy thinning, proper thinning has not been done in most of the teak monoculture plantations to avoid appreciable loss of increment. Moreover, pre-commercial thinning of teak plantation managed under PFP has been done without a scientific study (Jayarathne and Subasinghe, 2008). Thinning is often governed by revenuerather than silvicultural considerations. Ignoring possible difficulties in the future, thinning operations are delayed deliberately in order to harvest larger poles at a later stage. Eventually, in many monoculture plantations, teak does not tend to grow tall and straight.

Most Profitable Monoculture Plantation

Tree monoculture is generally practiced for economic advantages, including the economic advantages of specialization, and flexibility to intensively grow a high-value timber species ideally suited to local growing conditions. Teak in Sri Lanka is one of the most preferred species in order to make plantations economically viable. Teak wood is unique and finds application in diverse end uses in Sri Lanka. Teak is one of the most well known timbers of the world with qualities including attractiveness in colour and grain, durability, lightness with strength (Kaosa-ard, 1993). Teak plantations are used for industrial purposes and producing "high value" timber for industrial use. Teak wood produce in monoculture has considerable value-adding and income generating potential because quality of teak products is prescribed rigidly by the local and foreign markets. Teak growth in monoculture plantations is rapid and there is a relatively rapid turnover and produce homogeneous economic crops. Teak as a commercial wood, the price per unit is well fixed at different sizes and ages of the stand and at different points of being processed. Initial investments and their estimated returns for a teak plantation are given in Table – 2.

Table 1. Age class distribution and growing stock of Teak in Forest Department plantation

Age Class	Extent (ha)	Total Volume (m3)
0 ~ 5	8,580	
6 ~ 14	327	11,510
15	779	35,445
16 ~ 20	386	15,054
21 ~ 25	297	13,365
26 ~ 30	1,229	54,137
31 and 31+	14,735	559,930
Total	26,333	689,441

Source: (Ruwanpathirana, 2012).

Table 2. Investment and profits sheet for Teak (Prices are in Sri Lankan rupees)

Land ext	No. Of trees	Expected return, Rs	Investment, Rs
10 p	15	1,800,000.00	468,000.00
20 p	30	3,600,000.00	936,000.00
40 p	60	7,200,000.00	1,872,000.00
60 p	90	10,800,000.00	2,808,000.00
80 p	120	14,400,000.00	3,744,000.00
160 p	240	28,800,000.00	7,488,000.00

Source: Investment Plans - Sadaharitha Plantations Limited.
www.sadaharitha. com/contents/investment-plans.php.

Even though the reliable economic and financial appraisals are very limited, properly managed teak plantations in Sri Lanka could generate attractive returns to investment. According to the values predicted by the simulation model, the internal rate of return of teak monoculture plantation was above the market rate of 12% concerning the Teak Management Plan in 1997 (Gunaratne and Subramanium, 2007). Management of teak monoculture plantations is less complex than mixed planted forests. The teak trees grown in pure stand has high commercial value. According to a study, the mean dbh values (cm) of the teak timber species in monoculture plantation at harvestable age was 35.9, whilst the mean height values (m) was 18.6. Analyses revealed that the future values of a 1m3 of timber at the respective harvestable age would be Rs.260,000 for teak in 30 yrs. Based on these values, sawn wood and wood from thinning operations during the growing phase would fetch an income with a Net Present Value of Rs. 1216586 for teak. The Net Present Value of establishment and maintenance costs was Rs. 201524 for teak. The findings of the study have revealed that teak plantations are highly profitable compared to other timber species namely mahogany, E. microcorys and E. grandis (http://environmentlanka.com/blog/2009/feasibility-assessments-of-four-plantation-timber-species-2/.). An assessment of financial and economic feasibility of selected forest plantation species shows that Teak, Mahogany, Eucalyptus, Jak and Pine provide positive net present values at a 10% discount rate (Table – 3). Teak provides the highest NPV at the discount rate of 10% (116,635.63) and 15% (15,152.51) as well as the highest IRR value (16.52%) (Pitigala and Gunatilake, 2002).

Table 3. Financial NPV and IRR for different timber species

	NPV				IRR
	10%	15%	20%	30%	
Mahogany	89,533.58	(7,873.45)	(29,690.05)	(32,527.77)	14.18%
Teak	116,635.63	15,152.51	(21,117.55)	(37,116.22)	16.52%
Eucalyptus	110,999.96	8,259.82	(27,174.38)	(38,207.28)	15.73%
Jak	110,991.94	(3,097.51)	(31,728.43)	(38,654.83)	14.72%
Pine	3,883.12	(33,175.92)	(45,429.960)	(47,778.11)	10.34%
Pine-Resin	32,116.12	(19,700.89)	(38,622.82)	(45,772.04)	12.49%

Brackets indicate negative NPV.
Source:Pitigala and Gunatilake, 2002.

Table 4. Teak timber harvested by STC from FD plantation

	2008	2009	2010	2011	Average volume (m3/ha)
Harvested area (ha)	484	625	562	558.5	
Harvested volume (m3)	28,640	32,559	28,374	26,564	52

Source: (Ruwanpathirana, 2012).

Assure Adequate Supplies of Timber

Management attempts to optimize only one species which is characterized by fast growth and high productivity. Forest plantation monoculture also offers specific advantage as the means to assure adequate supplies of timber in Sri Lanka. A large share of the economic burden in forestry due to the greater demand for timber is being gradually shifted to teak plantations. Planting degraded land with teak is one of the suitable fast growing species yielding higher returns per unit area. The species composition of timber plantation area of particular timber has to be matched with demand at the market. State Timber Corporation produces adequate teak to meet the demand of Sri Lankan market. Production and Sale values of teak timber produced by State Timber Corporation have been recorded in 2011 as Log volume (30,074 m3), Average sale value of logs (Rs. 37,097/m3) Sawn timber volume (331m3) and Average sale value of sawn timber (Rs 78,986/m3) (Ruwanpathirana, 2012). Table 4 shows the quantity of timber harvested by State Timber Corporation (STC) from the total forest plantation under Forest Department.

Foreign Exchange Savings

Supplies of large amount of teak timber from in-country monoculture plantations have reduces timber imports and thus lead to foreign exchange savings. Sri Lanka spent Rs: 6,751,401,928 to import timber and timber base products in 2011. However, the country has earned Rs. 5,591,907,461 as foreign exchange by exporting the mainly the teak timber and teak timber base products. The foreign trade account deficit is Rs.1,363,343,295 for timber importation (Ruwanpathirana, 2012). However, expanded planting of teak trees without proper consideration of existing and projected demands as well as marketing arrangements, could result in a disastrous decline in prices and damaging timber production programs.

SOCIAL IMPACTS OF TEAK MONOCULTURE

Employment and Income Generation

Teak monoculture plantations in Sri Lanka involved with many labor-intensive activities including raising trees, periodic silvicultural practices, harvesting, transportation, and processing of timber. Those activities generate more employment and income compared to other forestry activities. The results of a study has revealed a poor growth rate and low financial benefits of teak plantation managed under PFP is due to the site quality and lack of proper silvicultural practices. The growth of teak plantation managed under PFP is poor compared with the Provisional Yield Tables. The financial benefits are predicted to be low as the Net Present Value will be USD 1,501 assuming the present discount rate as 10% (Jayarathne and Subasinghe, 2008).

Involvement of small holders in investment of teak, a fast growing species is a major way for income generation and abundant supply of timber to cater the increasing demand with continuing high man/ land ratio. FWLs serve as one of the most appropriate land use systems and strategies for rural development both in the short and long term programs especially when farmers are made to benefit in the final tree component of the woodlots. In FWLs, fields are being planted with crops and teak trees. Crops are harvested within a few years to produce timber for both nearby markets and national timber trade. Many farmers have their sights on greater returns from timber. In some cases, farmers who manage teak woodlots have to wait at least 20 years for final financial returns, given that there is little market for small wood from the first and second thinning. Farmers are highly resistance for pre-commercial thinning of established highly-stocked woodlots, about removing small trees in the absence of markets for small logs.

In comparison, the small-scale teak woodlots appear to contribute less to the national economy but provide more benefits to poor farmers. An increasingly aware market for green, clean products is emerging for art products such as carvings from teak hardwoods. However, local initiatives and investing in the development of forest-based industries and small-scale enterprises, such as sawmills, lumber yards, furniture and crafts production or the marketing of a range of forest products, are still lacking to offers a significant means to increase employment and household incomes at the verge of producing large amount of timber. Despite the importance of forest as an important source of many environmental and cultural needs, modern industrial forest plantation sector has been especially active in promoting monocultures as forestry placing a higher relative value on wood in national and international economic decisions. In the case of Sri Lanka, closed canopy natural forests which covered about 80% of the total land area of the country has been dwindled up to just over 20% at present. Hence, teak monoculture plantations have become important forest for solving many environmental issues providing alternative source of timber.

Limitation to Produce Food with Shrinking Terrestrial Resources

Agricultural diversification is needed to generate more sustainable pathways to equitable and healthy food production and access. Rural farmers often demand for land with the intention to increase food production and hence income for better standard of living. Monoculture forest plantations are critical to ensure the ability to produce food with ever-shrinking terrestrial resources. Commercial teak monoculture plantation establishments continue to increase in the future will further decrease farm sizes for food production and alternative livelihood opportunities in the rural communities.

Sri Lankan food is distinctive for their culture. Many traditional food items are obtained from shifting (Chena) cultivation. It is common critics that the establishment of teak plantations on degraded shifting cultivation lands generates negative impacts on local communities and their livelihoods. Long term agriculture practiced in traditional shifting cultivation land prevents the growth of natural vegetation in the fields due to severe soil degradation, and slowly converting them to bare land. Therefore, conversion of traditional chena lands to permanent agriculture fields is of doubtful benefit to the rural communities.

Elephants-Human Conflicts

Decline in forest cover over time has reduced the inhabitable area and resources for large populations of elephants. Elephant habitats continue to be degraded leading to habitat loss and fragmentation. Shifting (Chena) cultivation is generally considered a practice detrimental to the environment and a cause of habitat degradation. However, traditional shifting cultivation actually creates optimal habitat for elephants. Monoculture plantations have negatively impacted forest biodiversity leading to increased straying of wild animals. Establishment of monoculture teak plantations changing biologically diverse ecosystem has heightened levels of human-elephant conflict in Sri Lanka. Teak monoculture in plantations destroys the elephants' natural habitat suppressing nearly all undergrowth and causes soil erosion during the rainy season. In some cases, elephants strip off the bark of Teak trees. Teak has no deep stable anchoring root system and tends to topple when elephants rub their bodies against them. The communities living close to FWLs face increased Elephants damages due to development of their temptation for teak bark in FWLs (Nanayakkara 2001).Currently Chena cultivation is illegal, but it is found mostly in lands under the Forest Conservation Department. Chena lands, support very high densities of elephants, as the chena cultivation creates and maintains conditions ideal for wild elephants by providing a habitat with a high volume of food. Elephants are forced to raid human habitations and agricultural lands due to lack of fodder for wild herbivores under teak plantations. Elephants charter into human territory in the process of raiding agriculture as they see them as a source of food, led to crop damage, manslaughter and injuries to people. Many elephants that died over the recent past were due to retaliatory killings. Humans illegally killed elephants on many occasions in retaliation.

Environmental Impacts of Teak Monoculture

Conservation of Natural Forest

Meeting local, urban, and export demands for timber is one of the factors contributing to deforestation and forest degradation in Sri Lanka. Creating teak wood producing areas outside traditional forests attracted attention of the government as a viable strategy to arrest the forest loss. Teak monoculture plantations play an important role in alleviating pressure on natural forests for timber and fuel-wood production. Increased supply of teak timber from monoculture plantations have reduced pressure of timber extraction particularly the illicit felling from natural forests. Rapidly increasing demand for teak timber is making additional timber supply leading to little or no forest conservation. Monoculture industrial plantations of teak required large extent of land caused the clearance of natural forests in many places as negative environment impacts.

Fail in Protecting Biodiversity in Forest Ecosystems

Many forests vitally depend on many other species, including insects and birds that pollinate trees and feed on pests, and numerous microbial species that live on and in trees and that are especially critical to survival. Declining of populations of birds and beneficial insects affect the biodiversity in and near monoculture fields. Lack of biodiversity in monoculture fails to preserve genetic diversity in the plantations. Teak forest plantations in Sri Lanka have ignored the negative environmental effects and active in promoting monocultures at the expense of all other forest functions. Sri Lanka is considered as a biodiversity hotspot in the world. From 28.8% of total land or about 1,860,000 ha in Sri Lanka is covered by trees. However the country has 185,000 ha of planted monoculture tree plantations than primary forests, the most biodiverse and carbon-dense form of forest covering 9.0% or 167,000 ha. The high stocking density and lack of structural diversity of teak monoculture plantations are incapable of supporting biodiversity characteristics of native forests. Extensive teak plantation monocultures diminish biodiversity, the quality of water, and the structure and fertility of the soil resulting negative impacts upon the quality of local life. Early canopy closure leads to suppression of growth of under storey vegetation that need enough light to complete both their vegetative and reproductive life cycles.

Teak tree shed its leaves during the dry season reducing the amount of shading of the understory. Increased sunlight dries up water on the ground inhibiting the survival of flora and fauna of the understory. The broad leaves of teak plants which cover the ground have very low rate of decomposition. These leaves serve as mulch and prevent other plant species to grow under teak plantations. New forest will not be created on a teak monoculture plantation even after the harvesting of the teak trees for many years, without additional activities. However, teak plantations are managed in ways that enhance their biodiversity protection functions as important providers of ecosystem services such as maintaining nutrient capital, protecting watersheds and soil structure as well as storing carbon.

Table 5. Average soil erosion rates for different land uses

Land use	Soil erosion rate (t/ha)
Pine	10.0
Eucalyptus	12.0
Teak	8.0
Mahogany	8.0
Poor Quality plantations	15.0
Open land	25.0

Source: MFE 1995.

Reclamation of Degraded Lands

Many lands under shifting cultivation in Sri Lanka had been abandoned after few seasons due to irreversible loss of soil fertility. Secondary forest regeneration and succession, and the related restoration of soil fertility in the shifting cultivation lands are very slow. There was a great need for characterizing these soils for reforestation or re-vegetation for a long period before the introduction of FWLs. The farmers already have more productive alternatives to shifting cultivation in terms of the return to labor to plant teak and on a larger scale. FWLs are more responsive to environmental concerns in increasing the forest cover in scrub forests used for shifting cultivations for generations.

The teak plantations established on marginal or degraded lands, reclaim those land through organic matter accumulation, reduction in soil erosion, and rain water conservation. Teak based farmers woodlots provide an alternative use of land formerly used for shifting cultivation.

Establishment of teak plantations under farmer woodlots play positive role in the restoration of biodiversity to an impoverished landscape of degraded marginal lands. A considerable amount of nutrients are returned to the soil and are made available for re-absorption through litter-fall in teak plantations. The total annual litter fall of teak plantations has been found to be 11,255 kg ha-1. The total annual nutrients return from litter contributes a major share of 110.26, 17.50, 35.03, 348.97 and 78.46 kg ha-1 of N, P, K, Ca and Mg, respectively (Manivasakan et al., 2012).

Some farmers prefer teak monoculture woodlots because they do not expect agricultural crops from the degraded land; they have insufficient labour available to manage the land intensively; and they have extensive land as well as other sources of income that meet household needs. Moreover, teak monoculture on infertile soil, rocky soil or rocky terrain improves soil quality to some extent and prevents landslides or erosion.

Depletion of Soil and Soil Erosion

Intensive monoculture farming depletes soil and makes it vulnerable to erosion. Teak monoculture plantations are associated with the deterioration of soil and erosion when teak is planted on steep slopes where excessive suppression of undergrowth is taken place. Teak monoculture suppresses nearly all undergrowth and causes soil erosion during the rainy season. Promotion of teak monoculture at the expense of all other forest functions, the industrial plantation sector has to assess the negative environmental effects. Table 5 shows the average soil erosion rates is minimum in teak monoculture plantations (8 t/ha) compared to other different land uses.

Greater Risk under Changing Environmental Conditions

Monoculture teak plantations are at greater risk than genetically diverse stands when environmental conditions are changed. Cultivation of genetically diverse tree species can better survive under changing environments, because some tree species are vulnerable to certain changes while other tree species are thrived under changing environment. Monoculture teak plantations without genetic diversity are likely to increase contribute to greater instability of production under changing conditions. Many production failures caused by outbreaks of disease, invasions of insects, and climatic anomalies in teak monocultures are often reported in Sri Lanka. Teak monoculture plantations are highly susceptible for wind damages because teak belongs to the vegetation with surface, lateral rooting system and very high competition among teak stand. The pathogens and pests encounter less resistance to spreading in monoculture than they do in mixed stands. When the monocultures as host plants are more genetically uniform and crowded, pathogens spread more readily, and epidemics tend to be more severe. Defoliators and the skeletonizers attack teak plantations in young stage is a common issue in teak monoculture plantations in Sri Lanka (Subasinghe, 2007). Due to the long growth period involved with teak timber plantations, often pest attack are leading to mortality of trees. Teak monoculture plantations which alter natural balances are at much greater risk of catastrophic insect and disease losses than natural forests. *Hyblaeapuera*is the most widespread and serious pest among the defoliators also causes damage to teak in Sri Lanka (Tilakaratna, 1991). Better management including pest control determines the productivity of teak plantations. Timely pest control measures are easily adopted in teak monoculture as the defoliation being a visible process.

Face Risks of Fire

Monoculture timber plantations face risks of wild fire in Sri Lanka. Change from a biologically diverse ecosystem to a monoculture plantations resulted in major forest fires. Plantation forestry with even-aged teak monoculture in the dry zone in Sri Lanka is vulnerable for fire (Subasinghe, 2007). The slowly decomposing teak leaves which cover the ground of teak monoculture plantations dry up in the dry season and serve as fuel for wildfire. Increase of fire incidence in teak plantations due to accumulation of fuel, accelerate erosion and habitat destruction. The soil surface becomes exposed to the sun and rain after a fire outbreak results in erosion and loss of nutrients leading to soil degradation during the rainy season. Although the forest fires in Sri Lanka are of comparatively small in size, the environmental damage caused by forest fires is often much greater and takes many forms (Perera, 2001). People deliberately used fire in clearing forest for shifting cultivation in adjoining forest reserves and spread firs rapidly through teak plantations causing severe damages. Changing the temporary land preparation systems of slash and burn shifting cultivation to more permanent FWLs virtually stopped the frequent incidents of forest fire.Participatory management working circles with forest user groups have been formed under the forest department to prevent fire hazards in the FWLs especially comprising teak monoculture plantations (Ariyadasa, 2002).

C Sequestration and Contribution to C Trade

Teak based global sink of carbon management would certainly increase according to increased interest developed in the establishment of teak forest plantations during the past 20 years (Adalarasan et al., 2008). Teak monoculture plantations lead to forest conservation and therefore to C sequestration. The C sequestration of teak is sustainable because the high-value teak when harvested is converted into long-term carbon storage products. Teak keeps the carbon locked for years as one of the most durable wood species, and has an edge over other wood species which matching performance only after treatment. The total estimated carbon stocks of the existing forest plantations in Sri Lanka was amounted to 4.23 million metric tons in an area of 57618.8 ha in 2008 (De Costa and Suranga, 2012). A study conducted in Sri Lanka shows that the teak plantations can be effectively used as sinks for carbon storage up to 49 years. The mean carbon storage of teak plantations with rotation length of 50 years is 133.66 tC/ha (Abayasiri and Ranasinghe, 2000). Among the five tree species occupying the area, 89% of the total C stock contributes by monocultures while 21% is contributed by teak next to the *Pinuscaribaea* (44%). The average C stock of the teakis estimated as 42.70 t ha (De Costa and Suranga, 2012). According to the results of a recent study, carbon storage values for 10 year old Mahogany plantations is 242.0MT/ha and 9 year old Rambutan plantations is 3.8 MT/ha. The carbon values for 2 and 3 year old Sandalwood plantations are 0.1 and 0.6 MT/ha while those values for 2, 3, 4, 5, 6 and 7 year old Teak plantations are 4.9, 9.7, 13.1, 23.4, 29.7 and 50.1 MT/ha respectively. At the 20thyear, the carbon contents of Mahogany, Rambutan, Sandalwood and Teak plantations will be 242.0, 41.3, 69.5 and 326.7 MT/ha respectively which are equivalent to 887.2, 151.4, 254.78 and 1,197.7 MT/ha respectively(Subasinghe et al., 2015).

Monoculture commercial teak plantation has become a mitigation option lead to capacity-building through planning and implementation of mitigation projects as well as international negotiations in sharing GHG benefits. The trade-offs between two management options and ecosystem services from teak monoculture plantations in Sri Lanka can be linked to the concepts of REDD+. Sri Lankan teak monoculture plantations have the possibilities of financial gains through the mechanisms of United Nations Reducing Emissions from Deforestation and Degradation in Developing Countries Program. Using available vast areas of degraded forests, village commons, and farm lands for teak timber plantations is increasingly becoming an attractive C sequestration project. Carbon budgeting indicated that teak plantation is an accumulating system which considered as a mitigation option to reduce the increased atmospheric carbon dioxide level (Jha, 2003). Although the carbon in the soil is roughly equivalent or lesser than the above ground biomass due to degradation in the tropical forest, sustainable management of forest plantations will certainly enhance the soil carbon in the context of mitigating climate changes (Adalarasan et al., 2008). However, resources invested in monoculture plantations are withheld from forestry production models of Clean

Development Mechanism due to the negative impacts of monoculture tree plantations upon the social level.

CONCLUSION

Supplementary teak monoculture plantations have become a vital importance in Sri Lanka to produce renewable timber to satisfy the wood requirements and also to decrease pressure on natural forests. Small and medium scale teak monoculture plantation managed by farmers and communities make high social impacts particularly contributing income generation. Although the teak monoculture plantations do not severely affect the supply of food, aggravation of elephant-human conflict due to conversion of traditional shifting cultivation lands has become a very important social issue. Teak monoculture plantations make mixed impacts both positive and negative depending on the locations, management practices and perceptions. Teak plantation has become a climate change mitigation option and sharing GHG benefits. Establishment of teak monoculture plantations on depleted forest lands using proven land-use techniques and innovative modern methods would meet the greater demand for timber while sharing the economic burden in natural forest management. Hence, promoting of scientific information and developing research strategies on teak monoculture plantations and other alternatives has become vital for finding solutions for major constraints, the economic issues, social issues, and environmental issues.

REFERENCES

Abayasiri, S. &Ranasinghe, 0 M. S. H. K. (2000). The contribution of forest plantations (teak) in Sri Lanka in acting as a carbon sink to reduce greenhouse gases in the atmosphere. *Proceedings of the Sixth Annual Forestry and Environment Symposium 2000 of the Department of Forestry and Environmental ScienceUniversity of Sri Jayewardenepura, Sri Lanka.*

Adalarasan, R., Karikalan Mani, V. &Manivasakan, S. (2012). Carbon sequestration: estimation of carbon stock in teak (Tectonagrandis Linn. f.) ECOSYSTEM." *Proceedings of International Forestry and Environment Symposium, Colombo, Sri Lanka.* Vol. *12.* 2012.

Ariyadasa, K. P. (2002).the forest fire situation in Sri Lanka, IFFN No., 26 - January 2002, p. 96-100.

Ariyadasa, K. P. (2003). Teak in Sri Lanka: Resource Base, Issues and Challenges. Proceedings International Conference Quality Timber Products of Teak from Sustainable Forest Management 2-5 December 2003, Peechi, Kerala, India.

Asian Development Bank (ADB). (2003). Project performance audit report on the participatory forestry project (Loan 1183-SRI[SF]) in Sri Lanka, ADB, Manila, Philippines.

Centeno, J. C. (1997). The management of teak plantations. ITTO Tropical Forest Update, 7, 2. http://www.treemail.nl/teakscan.dal/files/mngteak.htm (accessed 4.08.2010).

Cook, R. J. & Weller, D. M. (2004).In Defense of Crop Monoculture. New directions for a diverse planet: Proceedings of the 4th International Crop Science Congress, Brisbane, Australia, 26 Sep - 1 Oct 2004 (accessed 11th May, 2015www.cropscience.org.au).

De Costa, W. A. J. M. &Suranga, H. R. (2012). Estimation of carbon stocks in the forest plantations of Sri Lanka, *Journal of the National Science Foundation of Sri Lanka*, 2012, *40* (1), 9-41.

De Zoysa, M. & Inoue, M. (2007). Community Forest Management in Sri Lanka: Concepts and Practices; Proceedings of International Symposium on "Forest Stewardship and Community Empowerment: Local Commons in Global Context. Kyoto, Japan, Oct 11 ~ 12, 2007.

Gunaratne, L. H. P.&Subramanium, T. N. (2006). Predicting the optimal rotation length of Teak plantations using a simulation model. *Proceedings of the International Forestry and Environment Symposium 2006 of the Department of Forestry and Environmental Science*, University of Sri Jayewardenepura, Sri Lanka.

Jayarathne, K. M. T. S. &Subasinghe, S. M. C. U. P.(2007). Evaluation of a Small Scale Teak Plantation Managed under the Participatory Forestry Programme*Proceedings of the 12th International Forestry and Environment Symposium*, 2007, Kalutara, Sri Lanka.

Jha, K. K. (2003). Storage and Flux of Organic Carbon in Tectonagrandis L. f. Plantations Raised in the Moist Deciduous Region. Proceedings International Conference Quality Timber Products of Teak from Sustainable Forest Management 2-5 December 2003, Peechi, Kerala, India.

Kadambi, K. (1972). Silviculture and Management of Teak" (1972). "Forestry Bulletin No. 24: *Forestry Bulletins* No. 1-25, 1957-1972. Book 23. (http://scholarworks.sfasu.edu/ forestrybulletins/23).

Kaosa-ard, A. (1993). Teak international provenance trials I. Wood production and quality. In Proceedings "50th Year of HuayTak Teak Plantation: Teak Seminar." 294-312 pp.

Kaosa-ard, A. (1999). Gain from provenance selection. Paper presented at the Regional Seminar on Site, Technology and Productivity of Teak Plantations, Chiang Mai, Thailand, 25-29 January, 1999.

Kendurkar, S. V. (2011). Micropropagation: An effective tool for mass production of quality planting stock of teakIn K. Jayaraman, & K. V. Bhat (Eds.), Innovations in the management of planted teak forests: Proceedings of the International Training Programme, Kerala Forest Research Institute, Peechi, India, 2011.

Kumar, B. M. (2011). Soil management in teak plantationsIn K. Jayaraman, & K. V. Bhat (Eds.), Innovations in the management of planted teak forests: Proceedings of the International Training Programme, Kerala Forest Research Institute, Peechi, India, 2011.

Manivasakan, S., Mani, A. K., Balasubramanian, A., Santhi, R.& Karikalan, V. (2012, April). Litterfall and Nutrient Accretions to Soil in Teak Plantation. In*Proceedings of International Forestry and Environment Symposium, Colombo, Sri Lanka*, (Vol. *12*).

Ministry of Forestry and Environment (MFE), (1995). Forestry Sector Master Plan, Ministry of Forestry and Environment, Battaramulla, Sri Lanka.

Mercer, D. E. Adoption of agroforestry innovations in the tropics: A review. In *New Vistas inAgroforestry*; Nair, P. K. R., Rao, M. R., Buck, L. E., Eds.; Kluwer Academic Publishers:Dordrecht, The Netherlands, 2004, pp. 311–328.

Nanayakkara, V. R. (2001). Regional study on forest policy and institutional reforms: Final report of the Sri Lanka case study; Asian Development Bank, Colombo, Sri Lanka.

Pancel L. (1993). Tropical Forestry Handbook, Springel-Verlag, Berlin, Germany, 809pp.

Perera, G. A. D. (2001). The secondary forest situation in Sri Lanka: A review; *Journal of Tropical Forest Science*, *13*(4), (2001), pp. 768-785.

Pitigala, G. H. &Gunatilake, H. M. (2002), An Assessment of Financial and Economic Feasibility of Selected Forest Plantation Species. *Sri Lankan Journal of Agricultural Economics*., Volume *4*, Part 1, pp. 121-135.

Ranatunga, M. S. (1972). *Notes on the planting of forest species*, *Sri Lanka Forester*., Vol. *10*, 3 & 4, 103-116.

Ruwanpathirana, N. D.(2012). Sustainable utilization of timber resources in Sri Lanka. Ministry of Environment, Sri Lanka, pp.56-70.

Sathurusinghe, S. A. A. (1998). Community participation in the production of fuelwood; in FAO (1998) *Woodfuel in Sri Lanka - Production and marketing*; FAO of The United Nations, Bangkok.

Skutsch, M. M. (1990). Social Forestry in Integrated Rural Development Planning Sri Lanka, Field Document No. 24, FAO Regional Wood Energy Development Programme in Asia GCP/RAS/131/NET Bangkok, Thailand.

Subasinghe, S. M. C. U. P. (2007). Plantation Forestry in Sri Lanka: Challenges and Constraints. *Proceedings Institute of Biology at Annual Sessions of the Institute of Biology*, Sri Lanka, 2007 September.

Tilakaratna, D. (1991). Parasites of the teak defoliator, *Hyblaeapuera. Sri Lanka Forester*, *20*, 23-25.

In: Monoculture Farming
Editors: Tapan Kumar Nath and Patrick O'Reilly

ISBN: 978-1-63485-166-4

Chapter 3

PLANTATION FORESTRY: PARADIGM TO MEET THE DEMAND OF THE FORESTRY RESOURCES IN BANGLADESH

Mohammed Kamal Hossain*
Institute of Forestry and Environmental Sciences,
Chittagong University, Chittagong, Bangladesh

ABSTRACT

The first attempt of raising forest plantations in the Chittagong Hill Tracts, Bangladesh was made in 1871 through importing seeds of teak (*Tectona grandis*) from Myanmar. Since then, plantation forestry has become a part of the overall clear felling followed by artificial regeneration with commercially valuable species in the hill forests management systems of the country. The natural multistoried forests of Bangladesh were under the process of conversion from natural to artificial through the removal of existing growing stock and planting with more valuable species and also with fast-growing and firewood species. Today of the total forest area, 84% (12, 04,000 ha) has been classified as natural forest and 16% (237,000 ha) as plantation forests. The plantation forests are composed of long rotation species (131,000 ha), short to medium rotation species (54,000 ha), mangroves (45,000 ha) and the rubber plantations (8,000 ha). To mitigate the demand of timber and fuel wood in Bangladesh, a priority program of introducing fast growing tree species was taken up. Extensive trials of *Pinus*, *Eucalyptus* and *Acacia* species were done to find out the very fast growing exotics for some difficult sites. After successive elimination, provenance, growth and yield trials, some promising species, e.g., *Eucalyptus camaldulensis* and *Acacia auriculiformis* were recommended for large scale afforestation/ reforestation programs especially with an intention of rapid greening the barren hills and marginal lands. Social forestry concept was getting priority with the involvement of local people and appears a successful program in many areas. However, the productivity of the present planting stock in Bangladesh is extremely low in comparison to regional and international standards. There are also reports that 20-30% of all plantations established during the last 40 years have been destroyed, and of those surviving, the stocking is much less than the expected standards. The plantations appear

* Corresponding author: E-mail: mkhossain2009@gmail.com, kamal.hossain@ifescu.ac.bd.

likely to produce well below the capacity of the site due to lack of quality planting materials and proper maintenance. All this has contributed to very low growth of 2.5 m^3 ha^{-1} yr^{-1} in areas where 7.5 m^3 ha^{-1} yr^{-1} was the original standard. The paper examines the pros and cons of plantation forestry in Bangladesh and the contribution of plantation forestry in conservation of the remaining natural forests of Bangladesh.

Keywords: plantation forest, fast growing, hill forest, productivity, coastal plantation

INTRODUCTION

The total area of Bangladesh is 14.757 million hectares, of which agricultural land consists of 65% of the geographic surface, forest lands account for almost 17.5% (2.53 million ha), while urban areas cover 8% of the land. Water and other land uses account for the remaining 10% (Alam 2008). Regarding the areas of forest land in Bangladesh, it vary considerably from one source to other, and there is a general want of reliable information in this regard, e.g., the total forest area is 2.56 million ha (FMP 1993), 2.53 million ha (Alam 2008), 2.25 million ha (BBS 1999) and 1.47 million ha (World Bank 1997). The total forestland includes classified and unclassified state lands, homestead forests and tea and rubber gardens. Of the 2.53 million ha of forest land, the Forest Department manages 1.53 million ha including hill, Sal and Sundarban forests (Alam 2008). The remaining 0.73 million hectare of land designated as Unclassed State Forest (USF) and are under the control of Deputy Commissioners of the hill districts (Table 1).

Table 1. Total forest land and forest area managed by Bangladesh Forest Department

Category of Forests	Area (million ha)	% of the total country
Forest Department Managed Forests	1.53	10.54
Unclassed State Forests (USF)	0.73	5.07
Village or Homestead Forests	0.27	1.88
Total	2.53	17.49

Forest Department Managed Forests	Area (million ha)	% of the total country
Hill Forests	0.67	4.65
Natural Mangrove Forests	0.60	4.09
Mangrove Plantations	0.14	0.97
Plain Land Sal Forests	0.12	0.83
Total	1.53	10.54

However, another survey of Bangladesh forests is a little bit different than the existing area information (Altrell et al., 2007). Of the total forest area of 1.44 million ha, 84% has been classified as natural forest and 16% as plantation forests (Table 2).

Village forests (homesteads) form the most productive tree and bamboo resources in the country and account for 0.27 million hectares. Much of the Government forest lands do not have satisfactory tree cover (World Bank 1997, Chowdhury 1999, Ahmed 2003) and only

0.84 million ha (about 5.8%) of the state forest land has good forest cover (FSP 2004). The hill forest and bamboo forest in general have a tree cover of 10-70% (Altrell et al., 2007). All the mangrove forest areas have a very high tree cover, whereas the plantations in general have a low tree cover.

Table 2. Total forest areas of Bangladesh by National Land Use Classes (Altrell et al., 2007)

Forest Area (ha)	Category (ha)	National Land Use Classes and Area (ha)		%
14,42,000	Natural forest 12,04,000 ha	Hill Forest	551,000	38.2
		Sal Forest	34,000	2.3
		Mangrove Forest (salt water)	4,36,000	30.2
		Bamboo or mixed Bamboo/broadleaved forest	184,000	12.7
	Forest plantations 237,000 ha	Long rotation forest plantation	131,000	9.1
		Short/medium rotation forest plantation	54,000	3.8
		Mangrove plantation	45,000	3.1
		Rubber plantation	8,000	0.5

Forests Types and Distribution in Bangladesh

The natural forests of Bangladesh consist of three major vegetation types occurring on three different land types (Hassan 1994). But, the forests are shrinking rapidly due to encroachments resulting from serious population pressure (Bhuiyan 1979, Hossain and Hossain 2014). Besides this, the management of these low yielding heterogeneous forests was very complicated. So, at the beginning of the management of these forests, the main objective was to replace these heterogeneous natural forests with the plantations of commercially valuable species (Hossain 1998).

Hill Forests

The hill forests of Bangladesh are ecologically divided in to tropical wet-evergreen and tropical semi-evergreen forests. The tropical wet evergreen forests are an important category of forests in Bangladesh in terms of biodiversity, forest resources and environmental concerns. These are magnificent dense evergreen forests rich in floral and faunal composition. The trees in the top canopy reach a height of 46-61 m (Das 1990). A few semi-evergreen or deciduous species may also occur but they do not really change the evergreen physiognomy of the forests. These forests occur usually in hills and moist shady areas in Chittagong Hill Tracts (Khagrachari, Rangamati and Bandarban), Sylhet, Habigonj, Moulvibazar, Cox's Bazar and Chittagong.

The tropical semi-evergreen type of forest occurs in more exposed dry locations of Cox's Bazar, Chittagong, Chittagong Hill Tracts and Sylhet. The top canopy species of the tropical semi-evergreen forests reach a height of 25-57 m. In this forest, the evergreen species predominate but also there are many deciduous species. Many of the evergreen species also occur in this type of forests.

Plain Land Sal (Shorea Robusta) Forests

The tropical moist deciduous forests are known as Sal forests. Theses occur in Dhaka, Gazipur, Tangail, Mymensingh, Sherpur, Jamalpur, Netrokona, Dinajpur, Rangpur, Naugaon, and in much degraded form in Comilla. In this forest, the predominant species is Sal (Shorea robusta). The trees are 10-25 m tall and mostly deciduous (Das 1990).

Mangrove Forests

Mangrove Forests are very important forest resources in Bangladesh. These are also called littoral swamp forests. These are mainly evergreen forests of varying density and height, always associated with wet soils. The mangrove forests are well developed in the Sundarbans (Khulna, Bagerhat and Satkhira) on the Ganges- Brahmaputra delta. However, the developed Chakoria Sundarbans are now degraded because of shrimp, salt and other land uses. The main species are *Heritiera fomes, Excoecaria agallocha, Avicennia officinalis* and many others constitute the tidal or mangrove vegetation.

The Sundarbans mangrove is now a World Heritage Site. The total area of the Sundarbans is 6,017 km^2, which is the single largest natural mangrove tract in the world. It is about 4.2 percent of the total land area of Bangladesh and about 44 percent of the forest land. About 3.5 million people are directly or indirectly dependent on the forestry resources of this Sundarban forests.

Plantation Forestry in Bangladesh

The first attempt of raising forest plantation in Bangladesh was made in 1871 with teak (*Tectona grandis*) in Chittagong Hill Tracts bringing the seeds from Myanmar. Since then plantation forestry has become a part of the overall clear-felling followed by artificial silvicultural system in the hill forests, but the plantation activities remained limited to the Chittagong Hill Tracts until 1920. In 1921, plantation programs were extended to the Cox's Bazar and Sylhet Forest Division. The total plantation area reaches to 4,140 ha in 1948 with an annual planting rate of 100 - 300 ha. Some important hardwood species, e.g., *Swietenia macrophylla, Lagerstroemia speciosa, Toona ciliata, Artocarpus chama, Xylia xylocarpa, Syzygium firmum* were introduced in the plantation programs along with *Tectona grandis* (Chowdhury 1982, Das 1982, Hossain 1998). In the 1950's and 1960's, sal (*Shorea robusta*) plantations were established over substantial areas (Table 3). In this way, the natural multistoried forests of the country were under the process of conversion from natural forests to plantations through the removal of existing growing stock and planting with more valuable species and plantation area gradually reached to 72,000 ha in 1968. However, plantations are increasing day by day to cover the barren and vacant forest lands as well to meet the need of forest resources of the huge population.

The Forest Department started planting fast-growing tree species, e.g., *Gmelina arborea, Paraserianthes falcataria* and *Anthocephalus chinensis* in 1974. The total plantation of the country in 1990 was 3,32,000 ha of which 113,000 ha were in coastal regions, 21,100 ha in the sal forest areas and the rest 198,000 ha in the hill forests. Most of the sal plantations were in non-existence and only 122,000 ha of other long rotation plantations are traceable (FMP 1993). The Forestry Master Plan (1993) suggested two scenarios to reach the country's forest area to 20% by 2012. The first one proposed was an annual planting target of about 18,000 ha during 1993-2002 and 21,000 ha during 2003-2012. The second projection was an annual planting target 17% higher, in order to attain sustainable management of forest resources of the country. A potential area of about 700,000 ha for industrial plantation programs exists in the Chittagong Hill Tracts, Cox's Bazar and Sylhet districts. Plantations established by the Forest Department between 1995-96 through 2012-2013 are shown in Table 4 (personal communication with the Forest Department).

Table 3. Tree species introduction in the plantation forests of Bangladesh

Year	**Species introduced in plantation programs**
1871	Introduction of Teak in Kaptai, Chittagong Hill Tracts bringing seeds from Myanmar.
1871-1920	Plantations limited in Chittagong Hill Tracts with teak.
1921	Plantation extends to Chittagong, Cox'sBazar and Sylhet Forest Divisions. *Dipterocarpus turbinatus, Swietenia macrophylla, Lagerstroemia speciosa, Toona ciliata, Artocarpus chama, Xylia xylocarpa* and *Syzygium firmum* are included in the plantations.
1923	*Gmelina arborea* also included in the plantations.
1950	Till teak dominates.
1950-1960	*Shorea robusta* plantation in substantial areas but in many areas failed.
1960	Initiatives of experimental rubber plantation.
1965	Coastal plantation establishment starts with *Sonneratia apetala, Avicennia officinalis* etc.
1971	Independence of Bangladesh.
1974	Large scale plantation of *G. arborea, Paraserianthes falcataria, Anthocephalus chinensis* in Chittagong, Chittagong Hill Tracts and Sylhet forest areas. • included Oil palm, Mulberry and Cashew nut in plantations. • BFIDC started commercial rubber plantations.
1977-78	Initiation of species and provenance trials of *Eucalyptus* and *Acacia*.
1980 and onwards	Long rotation (40 years) plantations of *T. grandis, D. turbinatus, S. firmum, A. chama, M. champaca, S. macrophylla, C. tabularis* etc. Medium – Short rotation (18, 12, 6 years): *E. camaldulensis, A. auriculiformis, A.* hybrid, *Albizia procera, Gmelina arborea, M. azedarach, Cassia siame*a etc.
1990	*Khaya anthotheca* (Lombo) was introduced from India by some nurserymen and planted haphazardly, especially in the north Bengal areas.
2008 and onwards	*Tectona grandis, Acacia auriculiformis, Gmelina arborea, Eucalyptus camaldulensis, Dipterocarpus turbinatus* are getting priority in plantation programs.

Coastal Plantations

Bangladesh is the pioneer in coastal afforestation programs (Hassan 2013). Establishment of coastal plantations was started in 1961 by the Forest Department with a view to providing protection against natural calamities in the coastal areas, such as cyclones and tidal surges. The Forest Department started intensive mangrove plantations from 1965 with the objectives of protection of coastal life and property from tidal surges and cyclones, stabilization of the fragile coastal land, and meeting demands for fuel wood and industrial raw materials. The man-made mangroves established by the Bangladesh Forest Department are in the districts of Patuakhali, Bhola, Noakhali, Chittagong and Cox's Bazar (Table 5). In addition, there are 10 ha coconut, 40 ha Areca and 280 ha of bamboo and cane plantations established in the coastal areas.

Table 4. Plantations established by the Forest Department during the last 18 years (1995-96 to 2012-2013)

Sl. No	Fiscal year	Area (ha) in woodlot, agroforestry, buffer zone etc.	Length (Km) in strip plantation, embankments etc.
1	1995-96	8574.85	1665.91
2	1996-97	8450.77	3953.09
3	1997-1998	8964.67	1719.00
4	1998-1999	11079.30	1563.00
5	1999-2000	11503.29	2255.00
6	2000-2001	16247.47	4042.50
7	2001-2002	7327.66	3898.21
8	2002-2003	13290.603	4223.93
9	2003-2004	20823.88	4801.72
10	2004-2005	14440.91	4145.44
11	2005-2006	18637.85	4397.00
12	2006-2007	5016.50	216.10
13	2007- 2008	6998.26	863.10
14	2008-2009	6744.47	164.60
15	2009-2010	8497.14	762.10
16	2010-2011	13572.50	5623.10
17	2011-2012	17249.16	4796.32
18	2012-2013	10855.90	777.00
Total		2,08,275.183	49,867.12

These mangrove plantations are sources of timber, poles and firewood to the local people. These are also acting as ideal filter of coastal environment, disaster protector, habitat and breeding ground for fishes and shrimps, grazing land for cattle, water buffaloes, habitat for wildlife and venues for ecotourism (Siddiqi 2001). Some of the afforested mangrove islands are so dense and green that looks as natural forests. Nijhum Dwip, Dhal Chara and Char Kukrimukri are some such islands. Spotted deer were released in these islands and their number has increased considerably over the years. Establishment of successful plantations and creation of good habitat for wildlife, some coastal plantation sites, e.g., "Nijhum Dwip,"

Char Kukrimukri were declared as protected areas in recent years. So far, 1200 km^2 of land accreted from Bay of Bengal has been added to mainland of the country through coastal afforestations (Anon 2015).

Table 5. Coastal plantations raised by Forest Department (FD) during 1965–2013 (Hassan 2013)

Forest Divisions	Mangrove species	Non-mangrove	Nypa plantation	Strip plantation (km)
Bhola Coastal	39,203.30	1075.10	271.38	1603.17
Patuakhali Coastal	26,607.00	546.50	2601.50	3426.90
Noakhali Coastal	77,684.95	2377.25	-	3593.84
Barisal Social	8.50	-	-	1911.77
Chittagong Coastal	48,891.41	4680.68	-	1406.45
Feni Social	491.46	10.0	-	185.0
Cox's Bazar	-	-	-	
	192,395.24	8689.53	2872.88	12127.13 km = 4,850.85 ha

Social Forestry Plantation programs

Deforestation is an immense problem in Bangladesh. Over a long period, much of the state forests (Forest Department managed) have been deforested and encroached by the organized land grabbers and landless poor people. Prevailing socio-political condition does not allow eviction of these encroachers. Therefore, to solve this problem, instead of evicting the encroachers, they are actively involved in tree planting activities with a first right use in the designated forest areas. Plantations in social forestry program are raised with the active participation of local poor people under a benefit –sharing agreement (Khan et al., 2004).

The marginal land, such as slopes of roads, embankments and railway lines has also been brought under participatory forestry for raising strip plantations. Participatory forestry started in Bangladesh in 1981 with the commencement of the "Development of Community Forestry Project." Subsequently, two more participatory forestry projects were implemented, namely the Thana Afforestation and Nursery Development project and the Coastal Greenbelt Project. A participatory forestry project named as the Extended Social Forestry Project was implemented during 1995-97. Under the social forestry projects, 45,830 ha of woodlot plantations, 10,626 ha of agroforestry and 62,329.38 km of strip plantations were established (Anon 2015).

The rotation of social forestry plantations was 8 to 10 years. Meanwhile, 24,184 ha of woodlot/agro-forestry plantations and 10,944 km of strip plantations have been harvested. The total revenue earned from these harvest was equal to Tk 528.93 crore. Participants share was Tk 237.26 crore and this money was distributed to 109,400 participants. The plantations harvested and the shares distributed to the participants from 1999-2000 to 2013-14 are as in the Table 6.

Table 6. Summary of harvested Social Forestry plantation (1999-00 to 2013-14)

Area felled (ha/ km)	Timber quantity (Thousand m3)	Fuelwood quantity (Thousand m3)	Poles (Thousand Nos)	Total sale Proceed (Crore Tk)	Participant Involved (Thousand Nos.)	Participant share (Crore Tk)	Tree Farming Fund (Crore Tk)	GOB Revenue (Crore Tk)
24,184 ha & 10,944 km	464.66	508.26	4775.93	528.93	109.40	237.26	51.97	220.36

Undoubtedly, participatory social forestry is helping to reduce poverty. A total of 0.38 million people are involved in participatory social forestry plantations in the country. As a result, members of these families are receiving substantial benefits, which is a big step in poverty reduction. For smooth implementation and sustainability of social forestry, the government of Bangladesh promulgated the Social Forestry Rules in 2004. This will institutionalize social forestry program throughout the country, which, in turn, will ultimately preserve environment and conserve biodiversity of the remnant natural forests.

In addition, the Forest Department, NGOs and other institutes or organizations are actively involved in plantation activities in the country. Some of the notable activities and initiatives are as follows:

- Participatory Natural Sal Coppice Forest Management: Sal (*Shorea robusta*) is the key tree species of plain land Sal forests. This forest is now under heavy population pressure as it is easily convertible into agricultural land. The Forest Department started participatory Sal coppice management from 2001 under the Forestry Sector Project. An area of 5,850 ha Sal forest has been brought under coppice management through participatory social forestry programs.
- Participatory Barind Ponds and Tanks Rehabilitation and plantation: Ponds and tanks of the Barind Tract in the northern region are important water sources for household purposes and irrigation. For this reason, the Forest Department has taken up re-excavation and plantation programs to conserve and develop these important water sources in collaboration with the local community.
- Participatory Buffer Zone Plantation: The Forest Department has taken a program to conserve and develop about 8,288 ha of national parks and wildlife sanctuaries with local community participation throughout the country. This program will help in conserving biodiversity, environment, soil, water and wildlife. The participatory buffer zone plantation is one of the major components of this program. These plantations will meet the local demand of forest products and also help reduce poverty.
- Livelihood Improvement of Tribal People through Participatory Forestry in Chittagong Hill Tracts: This is a new program and 124 families were involved under this program. Each household was provided with 2 ha of land where they developed Silvi-agri-horticultural farms. In addition, 820.0 ha of block plantations were raised

in highly degraded forest areas through people's participation in Lama and Bandarban Forest Division.

- Homestead Plantation: Homestead plantation is an important afforestation program of the government of Bangladesh. The main objectives of the program are to preserve and develop environment and protect homesteads from cyclones and tidal bores by creating windbreaks around homesteads. Under this program, fruit and timber seedlings were distributed at a subsidized rate to the rural households. A huge amount of tree resources was created at the rural household level. In addition to providing shelter for birds and environmental conservation, these are supplying families with food, necessary fuelwood and timber.
- Institutional Plantation: The government of Bangladesh has been implementing an institutional plantation program throughout the country with the objectives of re-greening public institutions and urban centers. Under this program, trees are planted in areas available within the various institutions, such as educational institutions - schools, colleges and universities, compounds of mosques, temples and churches, graveyards, office complexes and other institutions. Chittagong University campus is one of the successful plantation campus possessing more than 835 flowering plants (Momen et al., 2006). This institutional plantation programs are very helpful for preserving and developing the environment and also to increase the scenic beauty in the respective areas.
- Urban Plantation: Population in the urban areas increased tremendously in recent years, which is the major cause for disappearance of greenery in the cities and towns resulting in the degradation of urban environment. For this reason, the government started city plantation programs from 1992 under the Nagar Banayan project.
- Bamboo, Cane and Murta Plantation: Bamboo, cane and murta are important raw materials for cottage industries in Bangladesh. For this reason, the government raised participatory bamboo, cane and murta plantations under a development project. About 2,011 ha bamboo plantations, 3,991 ha of cane plantations and 784 ha of murta plantations were raised between 1989 and 2005.
- Medicinal Plant Plantation: The demand of herbal medicine has been increasing over the years. As the country possesses a huge diversity of medicinal plants, their conservation and promotion can contribute tremendously in meeting the huge demand of herbal medicine. The Government of Bangladesh laid emphasis on the afforestation and conservation of medicinal plants. So far, more than 300 ha area have been planted with medicinal plants.
- Neem Plantation: Neem (*Azadiracta indica*) is an environment –friendly plant which is rich with high medicinal and insecticidal values. The government has taken up a huge program for neem plantations.
- Coconut Plantation: Coconut plantation program was initiated in 2003 with the objective of achieving self-sufficiency in coconut oil and other products derived from coconut. The initial program was to plant 10 million coconut plants in the 26 coastal districts. So far more than 7 million saplings have been planted involving people all over the country.

- Agar plantation: The agar plant is the only raw material for perfume industries. The government established more than 8,000 ha of agar plantations in Sylhet, Chittagong, CHT and Cox's Bazar hill forests areas to feed the local perfume industries.
- Climate Resilient Participatory Afforestation and Reforestation Project (CRPARP): The project is implemented in 51 Upa-zilas in 9 coastal districts under 2 Divisions. The objectives of the project is to reduce forest degradation and increase forest coverage through participatory planning, monitoring and to contribute in building the long-term resilience in the selected coastal and hilly areas. It will include a total of 17,000 ha of land (both coastal and mainland) and 1,672 km of roadside plantations.

In Bangladesh, Tree plantation has now become a social movement. The Government of Bangladesh started the tree planting movement in 1991 with an aim to accelerate tree plantation activities throughout the country. The aim of the program was to create awareness about environment and nature conservation, to highlight the role of trees in environmental preservation and also to disseminate information on the importance of tree resources. The government also introduced the tree fair as a part of the "Tree Planting Movement in 1994. The main objectives of the tree fair are to disseminate new planting technology; to provide planting materials of high yielding varieties; to create awareness about tree planting and preservation of environment and to acquaint people about plant biodiversity. The Prime Minister of Bangladesh inaugurates the National Tree Fair on 5th June in Dhaka every year. During the monsoon, tree fairs are also held in the divisions, districts and sub-districts (upa-zila) all over Bangladesh. Considering the importance of tree planting for environmental and biodiversity conservation in Bangladesh, the Government of Bangladesh introduced national awards of the Prime Minister for outstanding performance in tree planting activities since 1992.

POTENTIAL SPECIES FOR PLANTATION PROGRAMS

The plantations have some advantages over natural forest from management and economic point of views, such as more production and ability to choose species with desirable characteristics. Plantation programs are getting priority in both public and private sectors of the country. But, the choice of species is still under debate, particularly regarding the plantations of some exotics like *Eucalyptus, Acacia* and *Pinus* species. Teak was the first species largely used in plantation forests followed by *Swietenia macrophylla* in Chittagong and Chittagong Hill Tracts. From 1968, elimination, provenance, growth and pilot plantation trials were established with a number of indigenous and fast growing exotic tree species in different silvicultural research stations of Bangladesh Forest Research Institute (BFRI).

In 1974, the Forest Department began establishing plantations of fast growing species such as *Gmelina arborea, Paraserianthes falcataria* and *Anthocephalus chinensis* in Chittagong Hill Tracts and Sylhet Forest Divisions. In 1978 and onwards, extensive trials of *Eucalyptus* and *Acacia* species were started to find out the very fast growing exotics for some degraded sites. *Eucalyptus camaldulensis, E. tereticornis* and *E. brassiana* emerged as the promising species for planting in Bangladesh from the trials of 84 seedlots of 12 species of *Eucalyptus* imported from different parts of Australia (Davidson and Das 1985). Other provenance trials of *E. brassiana* (12 provenances) and *E. urophylla* (12 provenances) were laid out in 1983 to select the superior provenances for Bangladesh (Hossain et al., 1996a, Islam et al., 1997). However, the seeds of a total of 216 provenances of 37 species of *Eucalyptus* were introduced in Bangladesh from which finally *E. camaldulensis*, petford-1 provenance was found suitable, promising and recommended for plantations (Davidson and Das 1985, Hossain et al., 1989). A comparative study of the coppicing ability of different provenances of *E. camaldulensis, E. tereticornis* and *E. brassiana* showed that their coppicing ability was very high and *E. camaldulensis* showed the highest height and diameter (dbh) growth of their coppices (Hossain et al., 1994). Hossain et al., (1996b) also reports the superiority of *E. camaldulensis* in Chittagong University hilly areas over other eucalypt species and provenances.

Among the *Acacia* species introduced so far in Bangladesh, *Acacia mangium* and *A. auriculiformis* have shown promising growth in poor afforestation sites (Latif et al., 1985, Hossain et al., 1997, Hossain et al., 1998a). Provenance trials of *Pinus caribaea* var. *hondurensis* were established in 1983 and 1984 at Keochia, Chittagong and found that provenances grown well in the area (Latif and Zashimuddin 1994) and *Pinus caribaea* var. *hondurensis* is doing better in comparison to *Pinus oocarpa* in the same location (Zashimuddin et al., 1991).

Another elimination trial of 23 dry zone hard wood species from Central America was laid out at Charkai (Dinajpur) in 1986 but the trial was failed at the initial stage of the establishment (Hossain et al., 1989). Similarly, a provenance trial of *Acacia mangium* with 23 provenances (19 from Australia, two from Indonesia and two from Papua New Guinea) were established at Keochia, Chittagong (Zashimuddin et al., 1983). Records reveal that more than 130 exotic tree species so far tried in the plantation programs of Bangladesh (Hossain et al., 1989), but only a few of them proved successful in large scale plantation programs. Some of the successful exotic species and their status in the plantation programs are shown in Table 7.

Successful plantation technologies have been developed in many tropical countries of the world (Evans 1992, Kanowski and Savill 1992). Light demanding, colonizing exotic species have been the most successful in monocultures under plantation management programs (Hughes 1994). The tropical and sub-tropical plantation forestry has focused on a small number of fast growing, colonizing species, e.g., *Acacia, Eucalyptus, Gmelina, Pinus, Populus* and *Tectona* (Evans 1992, Harwood and Nambiar 2014). These species have the ability to capture the site rapidly and tolerate harsh soil and climatic conditions and protected from animals, humans and fire etc. These characteristics of exotics are a pre-requisite for success on the often highly degraded sites. Higher growth and yield of exotics over indigenous species have been attributed because of their greater tolerance of degraded sites and their escape from specialized pests and diseases. Thus, the diminishing natural forest resources are being compensated by rapid expansion of planted exotic trees worldwide (Davidson 1995). Successful exotics are being planted in some other countries also, such as Spruce in Europe, Radiata pine in Australia and New Zealand, Ipil-ipil in Philippines and Australia, Eucalypts in India, China and Brazil (Evans 1992).

In Bangladesh, afforestation programs with exotics also helps in improving the socio-economic condition of the rural people by generating employment in raising nursery and plantation activities (Hossain 1998). Some exotics in the social forestry programs are able to provide returns very shortly that encouraged the farmers in involving with the social forestry programs. Above all the pioneer colonizing species, e.g., *A. auriculiformis* are able to create a vegetation cover in degraded forest areas. Similarly, *Acacia auriculiformis, A. mangium* and *E. camaldulensis* proved superior in comparison to some native tree species at Madhopur sal forest areas of Bangladesh (Hossain et al., 1995).

Table 7. Exotics frequently used in plantation forests of Bangladesh

Exotic tree species	Origin	Status in plantations
Acacia auriculiformis	Australia	Extensively used in plantations all over the country.
Acacia mangium	Australia/PNG	Promising species but discourage due to heart rot disease infestation.
Eucalyptus camaldulensis	Australia	Government has ban on further planting of the species but small growers and individuals prefer the species very much in their homesteads and marginal lands.
Eucalyptus tereticornis	Australia	Limited plantation often mixed with *E. camaldulensis* plantations.
Hevea brasiliensis	Brazil/Malaysia	Rubber estates are commercially developing in hill forests and sal forest areas.
Khaya anthotheca	India	Widespread plantation in roadside, institution and homesteads by individuals, NGOs for its early promising growth, but diameter growth is very slow and the wood is also not promising.
Leucaena leucocephala	Central America	Planted in plain land and coastal areas.
Melaleuca leucadendra	Australia	Limited plantations.
Paraserianthes falcataria	Java	Planted as fast growing soft wood species.
Pinus caribaea	Honduras	Limited in small scale plantations.
Pinus oocarpa	Honduras	Limited in small scale plantations.
Samanea saman	West Indies	Naturalized, mostly planted in roadside and homesteads.
Swietenia macrophylla	West Indies	Extensively planted in forest, marginal land, institutes and homesteads.
Tectona grandis	Myanmar	Naturalized and dominant species in plantation forests.
Xylia xylocarpa	Myanmar	Promising species but limited plantations.

However, there is an increasing concern among foresters, ecologists, botanists, conservationists and policy makers about the threat of uncontrolled introduction of aggressive tree species in the plantation programs. Invasion of exotics may cause major loss of biodiversity and species extinction either due to direct replacement by exotics or indirect effects on the ecosystem. Concern also exists on the degradation of the environment, e.g., the controversial effect of eucalypts on environment. There are also risks of the decline of growth and yield in second and successive rotations, or the infestation of pests and diseases, e.g.,

Psyllid in *Leucaena leucocephalla*, shoot borer in *Swietenia* and leaf defoliator in *Tectona grandis* (Evans 1992, Harwood and Nambiar 2014).

WAY FORWARD

In Bangladesh, plantation program is increasing day by day and the exotic species are getting preferences over the indigenous ones in some degraded forest areas (Ara et al., 1989, Amin et al., 1995). With deforestation, many degraded sites are available for plantation, though very little is known about how to manage such sites economically. Conversion of healthy prospective natural forests to plantations of exotic species must be strongly discouraged. Productive plantations on previously deforested or non-forested sites may offer important options for conservation of natural forests and arresting the further deforestation activities. Another option of mixed species plantations of both the indigenous and exotics may also be able to contribute as an effective mechanism for sustainable resource management. Through the increase of productivity, pressures can be reduced on remaining natural forests, but emphasis must focus on existing plantations, and new plantations on degraded or denuded land, agroforestry and community forestry programs rather than on the existing natural forests.

The wastelands and denuded hillocks adjacent to human habitation which are either covered with sungrass or overgrazed by leaving the sites with a thin vegetation cover and of very little water retention capacity may be reforested with exotics like *Acacia* and *Eucalyptus* species. When the techniques of successful teak, eucalypts and *acacia* plantations were being at standardized, the environmentalists became conscious of the danger of raising plantations of these species. Considering the demand of the peoples, artificial plantation programs may support but the major questions still concern is the correct choice of species for the site - whether mixed or in pure stands and the interactions of species and site, especially where exotic species like *Eucalyptus* and *Acacia* are used.

Table 8. The Potential land available for social forestry practices in Bangladesh

No.	Land availability	Area (million ha)
01	Degraded & denuded land of Unclassed State Forest Land	1.00
02	Khas lands	0.56
03	Degraded government forest land	0.27
04	Marginal strip land	0.08
05	Homestead marginal land	0.27
06	Degraded tea garden land	0.06
07	Degraded private forest land	0.05
08	Cropland Agroforestry on private agricultural lands (29% of the total agricultural land is above normal flood level & suitable for cropland agroforestry)	2.36
	Total	4.65

Bangladesh has an immense scope of popularizing social forestry programs in different land components. In total about 4.65 million ha of land is available for some sorts of social forestry programs, which may be about 31% of the country's total land surface (Table 8). Considering the size of Bangladesh and her forest area, the potential land available for social forestry production systems is quite significant (Khan et al., 2004, Ali and Morshed 2011).

If all the available 4.65 million ha of land is brought under social forestry production systems then 465,000 ha of land would be available for planting annually under the social forestry production system of 10 years rotation. The area will produce 46.5 million m^3 of timber and firewood annually (at the moderate rate of 10 m^3 ha^{-1} yr^{-1} production). Moreover the system will provide food, income and employment opportunities for the farmers and participants of the programs.

CONCLUSION

The natural forests of Bangladesh are under severe pressure for meeting her growing demands for timber, fuel, fodder and non-timber forest products from an ever increasing human population, and industrial demands of the country. The annual estimated demand of round logs in the country is 15.3 million m^3 and supply is 8.3 million m^3 (Ahmed and Akhtaruzzaman 2011). There is a gap of 7.0 million m^3 between demand and supply and the gap is increasing with time. There is a moratorium on harvesting of forest produces (timber) since 1989 and shall continue till 2015 (Hossain et al., 2008). To meet the requirement, Bangladesh is importing timber from abroad. Since last twenty four years (1989-90 to 2012-13) about 2.43 million m^3 timbers were imported in Bangladesh (personal communication with DFO, Chittagong North Forest Division) and the import is increasing day by day to meet the huge demand of the large populations.

The Government of Bangladesh has given top priority on tree planting in the degraded forests to re-green them through a benefit –sharing mechanism with local people's active participation. Participatory forestry is being recognized increasingly as the only way to re-green the degraded forests in Bangladesh. Through 30 years of extension work in participatory forestry and national and regional tree fairs all over the country, tree planting has now become a social movement for greening the country as well as to improve the environment. Collaboration among government organizations, NGOs, local government bodies and people in general is essential to move forward this movement to a success.

The annual estimated demand of round logs in the country is 15.3 million m^3 but the supply is 8.3 million m^3. There is a gap of 7.0 million m^3 between demand and supply and the gap is increasing with time. How to overcome the problem? Instead of spending the hard earn foreign currency for importing timbers; it will better to invest the money for plantations. Present plantation establishment cost is about 47,000 taka (US $ = 78 taka) per ha including 3 years maintenance cost. If we are able to spend the money for plantations, we may convert the all available vacant and barren lands to plantations easily.

REFERENCES

Ahmed, F. U. and Akhtaruzzaman, A. F. M. 2010. Agricultural Research Priority: Vision 2030 and beyond sub-sector: Forestry. Bangladesh Forest Research Institute and Bangladesh Agricultural Research Council (BARC), Farmgate, Dhaka, Bangladesh.

Ahmed, I. U. 2003. Forest Management: Bangladesh Perspective (In Bengali), *Aranya* 11(6): 19-28.

Alam, M. K. 2008. Forests and Forestry in Bangladesh. *In*: Ahmed, Z. U., Begum, Z. N. T., Hassan, M. A., Khondker, M., Kabir, S. M. H., Ahmad, M., Ahmed, A. T. A., Rahman, A. K. A. and Haque, E. U. (eds.), Encyclopedia of Flora and Fauna of Bangladesh, Vol. 1. Bangladesh profile, Asiatic Society of Bangladesh, Dhaka. pp. 73-86.

Ali, M. Y. and Morshed, H. M. 2011. Towards a new forest policy in Bangladesh. *In:* Proceedings First Bangladesh Forestry Congress 2011, pp. 360-378.

Altrell, D., Saket, M., Lyckeback, L., Piazza, M., Ahmad, I. U., Banik, H,, Hossain, M. A. A. and Chowdhury, R. M. 2007. National Forest and Tree Resources Assessment 2005-2007. Bangladesh Forest Department, Ministry of Environment and Forest; Bangladesh Space Research and Remote Sensing Organization, Ministry of Defence, Food and Agriculture Organization of the United Nations. pp. 192.

Anonymous. 2015. Forest Department Information Booklet. 26 pp.

Amin, S. M. R., Ali, M. O. and Fattah, M. I. M. (eds). 1995. Eucalypts in Bangladesh. Proceedings of a seminar held at the Bangladesh Agricultural Research Council, Dhaka on April 6, 1994. 73 pp.

Ara, S., Gafur. M. A. and Islam, K. R. 1989. Growth and biomass production performance of *Acacia auriculiformis* and *Eucalyptus camaldulensis* reforested in the denuded hilly lands. *Bangladesh J. Bot.* 18(2):187-195.

BBS. 1999. Statistical Yearbook of Bangladesh. Bangladesh Bureau of Statistics, Ministry of Planning, Govt. of the Peoples Republic of Bangladesh.

Bhuiyan, A. A. 1979. Afforestation of unclassed state forests lands areas in Chittagong Hill Tracts of Bangladesh. *Bano Biggyan Patrika,* 8 (1&2):64-70.

Choudhury, R. A. 1982. Afforestation of denuded hills of Chittagong and Chittagong Hill Tracts. Proceedings of Second Bangladesh National Conference on Forestry, Dhaka. pp. 1-5.

Chowdhury, Q. I. 1999. Bangladesh: Country Overview. *In:* Chowdhury (ed.), *Bangladesh State of Environment Report 1999*, Forum of Environmental Journalists of Bangladesh, Dhaka, pp. 1-14.

Das, S. 1982. The forest management practice in Bangladesh. Proceedings of Second Bangladesh National Conference on Forestry, Dhaka. pp. 6-18.

Das, D. K. 1990. *Forest Types of Bangladesh.* Bulletin 6, Plant Taxonomy Series, Bangladesh Forest Research Institute, Chittagong. 9 pp.

Davidson, J. 1995. Ecological aspects of *Eucalyptus* plantations. Proceedings of the Regional Expert Consultation on *Eucalyptus*, 4-8 October, 1993, vol. 1, RAPA publication:1995/6. FAO Regional Office for Asia and the Pacific, Bangkok, pp. 35-72.

Davidson, J. and Das, S. 1985. *Eucalypts in Bangladesh-A review*. Bangladesh Forest Research Institute, Chittagong, Bangladesh. 246pp.

Evans, J. 1992. Plantation Forestry in Tropics, 2nd ed., Clarendon press, Oxford. 403pp.

FMP (Forestry Master Plan). 1993. *Forestry Master Plan: Main Plan-93/2012, vol.1*, Ministry of Environment and Forest, Govt. of Bangladesh. 162 pp.

FSP. 2004. Participatory Forestry Newsletter, Bulletin No. 1. May 2003. Forestry Sector Project, Forest Department, Dhaka.

Harwood, C. E. and Nambiar, E. K. S. 2014. *Sustainable plantation forestry in South- East Asia*. ACIAR Technical Reports No. 84. Australian Centre for International Agricultural Research: Canberra. 100 pp.

Hassan, M. M. 1994. Forest soils of Bangladesh. *Bangladesh J. For. Sci.,* 23(2):1-11.

Hassan, D. Z. 2013. Plants in Mangroves & Coastal Afforestation in Bangladesh. 164 pp.

Hossain, M. K., Islam, Q. N., Islam, S. A., Tarafdar, M. A., Zashimuddin, M. and Ahmed, M. 1989. Assistance to the Second Agricultural Research Project, Bangladesh. Status Report on the Activities of the Silvicultural Research Division, BFRI, FAO, Chittagong. 20 pp.

Hossain, M. K., Zashimuddin, M., Islam, S. A., Tarafdar, M. A. and Islam, Q. N. 1994. Growth and yield of coppice crops of three *Eucalyptus* species in the sal forests of Bangladesh. *Bangladesh J. For. Sci*., 23(2):61-66.

Hossain, M. K., Islam, S. A. and Islam, Q. N. 1995. Selection of plantation tree species for degraded sal forest areas of Bangladesh. *Bangladesh J. For. Sci*., 24(2):5-10.

Hossain, M. K., Zashimuddin, M., Tarafdar, M. A., Islam, S. A. and Islam, Q. N. 1996a. Seven year growth and biomass production of *Eucalyptus brassiana* provenances at Tangail, Bangladesh. *Bangladesh J. For. Sci.,* 25(1&2):43-48.

Hossain, M. K., Sarker, S. and Kamaluddin, M. 1996b. Initial growth performance of *Eucalyptus* species and provenances at Chittagong University Campus. Chittagong Univ. Stud., part II, Science, 20(1):129-131.

Hossain, M. K., Islam, S. A., Zashimuddin, M., Tarafdar, M. A. and Islam, Q. N. 1997. Growth and biomass production of some *Acacia* and *Eucalyptus* species in degenerated sal forest areas of Bangladesh. *The Indian For.,* 123(3):211-217.

Hossain, M. K. 1998. The potential role of plantations on the rehabilitation of native forest biodiversity in degraded hilly areas of Bangladesh. Proceedings 20th *Bangladesh Sci. Conf.* Part 2:99-104.

Hossain, M. K., Hossain, M. S. and Aryal, U. K. 1998. Initial growth performance of eight provenances of *Acacia auriculiformis* at Chittagong University Campus, Bangladesh. *The Indian For.,* 124(4):256-258.

Hossain, M.K., Alam, M.K. and Danesh Miah, Md. 2008. Forest restoration and rehabilitation in Bangladesh. *In:* Keep Asia Green Volume III "South Asia," Don Koo Lee (ed.), IUFRO World Series Volume 20-III, Vienna, pp. 21-65.

Hossain, M. K. and Hossain, M. A. 2014. *Biodiversity of Chunati Wildlife Sanctuary: Flora.* Arannayk Foundation and Bangladesh Forest Department. Dhaka, Bangladesh. pp. v + 175.

Hughes, C. E. 1994. Risks of species introductions in tropical forestry. *Comm. For. Rev.,* 73(4):243-252.

Islam, Q. N, Islam, S. A., Tarafdar, M. A. and Hossain, M. K. 1997. Growth and biomass estimation of *Eucalyptus urophylla* provenance trial in degenerated sal forest areas of Bangladesh. *Indian J. Forestry,* 20(1):78-81.

Kanowski, P. J. and Savill, P. S. 1992. Forest plantations: towards sustainable practice. Chapter 6, In: C. Sargent and S. Bass (eds.), Plantation Politics: Forest Plantations in Development. Earthscan, London. pp. 121-151.

Khan, N. A., Choudhury, J. K., Huda, K. S. and Mondal, M. I. 2004. *An overview of Social Forestry in Bangladesh*. Forestry Sector Project, Bangladesh Forest Department, pp. 198.

Latif, M. A. and Zashimuddin, M. 1994. Performance of different provenances of *Pinus caribaea* var. *hondurensis* in Bangladesh. *Bangladesh J. For. Sci.*, 23(2):26-31.

Latif, M. A., Khan, S. A. and Kamaluddin, M. 1985. Prospects of *Acacia mangium* for afforesttaion in Bangladesh. *Pakistan J. For.*, 35(1):7-12.

Momen, R. U., Hossain, M. K., Huda, S. M. S. and Khan, B. M. 2006. Assessment of plant diversity in Chittagong University campus. *Journal of Forestry & Environment*, 4:27-52.

Siddiqi, N. A. 2001. *Mangrove Forestry in Bangladesh.* Institute of Forestry and Environmental Sciences, Chittagong University, Chittagong. 201 pp.

World Bank. 1997. *Review of the Key Environmental Issues in Bangladesh*, Workshop Discussion Draft (May 1997), World Bank, Dhaka. pp 22.

Zashimuddin, M., Latif, M. A., Khan, S. A. and Davidson, J. 1983. Performance of different provenances of *Acacia mangium* in Bangladesh. *Bano Biggyan Patrika,* 12(1&2):57-61.

Zashimuddin, M., Latif, M. A. and Ahmed, M. 1991. Performance of different provenances of *Pinus oocarpa* in Bangladesh. *Indian J. Forestry,* 14(3):232-236.

In: Monoculture Farming
Editors: Tapan Kumar Nath and Patrick O'Reilly

ISBN: 978-1-63485-166-4

Chapter 4

A REVIEW ON THE ENVIRONMENTAL IMPACTS OF OIL PALM EXPANSION IN THE DEVELOPING TROPICS

John Garcia-Ulloa[1] and Janice Ser Huay Lee[2,*]

[1]Department of Environmental Systems Science, ETH Zurich, Zurich, Switzerland

[2]Woodrow Wilson School of Public and International Affairs, Princeton University, Robertson Hall, Princeton, NJ, US

ABSTRACT

Palm oil is a ubiquitous ingredient found in a wide range of consumer products, including cosmetics, food and beverages, and as a substitute for fossil fuels. The rapid expansion of oil palm in the last few decades has resulted in widespread deforestation, biodiversity loss, and carbon emissions in the tropics. It has also negatively affected the livelihoods of many local communities. However, oil palm production is a highly lucrative business and the oil palm industry has brought about positive impacts on rural development, regional economies as well as lifting small farmers out of poverty. Here, in this chapter we pull together the literature on the broad environmental impacts of oil palm expansion and present an overview of the most recent research which are related to the loss of natural habitats, biodiversity, and ecosystem services. We take into account geographical and managements contexts, and analyse how current sustainability initiatives are addressing these issues. Our analysis shows that the most developed oil palm sustainability initiative, a voluntary certification scheme known as RSPO, addresses to different degrees the majority of the impacts identified from the expansion of this crop at the plantation level. However, because of the broad range of requirements, the uptake of this certification scheme among smaller producers has been low. Furthermore, certification schemes have a limited role to tackle the broader impacts of oil-palm agriculture at larger scales. We argue that additional mechanisms that promote an integrated land-use planning approach to oil palm landscapes and producing regions are needed, if the negative cumulative impacts from these initiatives are to be minimized and the positive benefits maximized.

* Corresponding author: janice.jlsh@gmail.com.

Keywords: biodiversity, deforestation, *Elaeis guineensis*, ecosystem services, sustainable certification

General Introduction

Palm oil is a versatile vegetable oil which can be found in a range of products, from household items such as shampoo and soap, cosmetics, as well as processed food and beverages (Sheil et al., 2009). It is used widely in the oleochemical industry and has been marketed as a biofuel to reduce the automotive industry's reliance on fossil fuel (Pinzi et al., 2009). The demand for palm oil has risen steadily over the last few decades as a result of rising consumer demand (Corley 2009; Koh and Lee 2012). Global production of palm oil increased from ~1.5 Mt in 1961 to ~54.4 Mt in 2013, an almost 35 times increase in volume over four decades (FAOSTAT 2013). With increasing demands and a rising population, projected levels of palm oil demand are estimated to reach 120 to 156 Mt by 2050 (Corley 2009).

Palm oil is derived from the oil palm tree *Elaeis guineensis* which is a native plant from West and Central Africa (Corley and Tinker 2003). The oil palm is a perennial crop with a lifespan of approximately 25 to 30 years, of which its first 4 years are immature and do not produce fruit. The fresh fruit bunches (FFB), which are the fruits of the oil palm tree, are processed in a mill to extract crude palm oil (CPO) (Corley and Tinker 2003). On average, the oil extraction ratio for a typical mill is 20%, which is equivalent to 200 kg of CPO extracted from 1 ton of FFB (Corley and Tinker 2003). The oil palm has the highest yield of any oil seed crop (e.g., soybean, rapeseed), averaging 3-4 tonnes of vegetable oil per hectare per year in major palm oil producing countries compared to soybean or rapeseed which produces 0.37 to 0.75 tonnes per hectare (World Bank 2011) and also uses far less inputs (Guardian Sustainable Business 2014). The oil palm tree grows well under tropical climatic conditions with a temperature range of 24 – 30 C, and rainfall conditions from 1780 – 2280 mm (Sheil et al., 2009). Oil palm plantations are often monoculture plantations and are shade intolerant (Corley and Tinker 2003). As a result, they do not grow well as an agroforest crop such as coffee or cacao (Phalan et al., 2009). Based on FAO statistics for 2013, oil palm is cultivated in 44 countries (Table 1) and accounts for almost 18.1 M ha of cultivated land area in the world or ~0.02% of the world's permanent cropland (1.5 billion ha) (FAO 2015). Of the global land area of oil palm cultivation, 64% (11.6 M ha) occurs in the region of Southeast Asia, and more specifically within Indonesia and Malaysia.

Monoculture oil palm plantations have expanded rapidly since the 1960s, and most of the expansion has occurred in tropical Southeast Asia, dominated mainly by Malaysia and Indonesia. Between 2000 and 2013, the area of cultivated oil palm land increased from 3.07 M ha to 4.55 M ha (48%) in Malaysia, and 2.01 M ha to 7.08 M ha (252%) in Indonesia (Figure 1). With rising demands for palm oil, monoculture oil palm plantations are expected to expand especially in developing tropical countries across the world. In Africa, oil palm plantation land area has increased by 10% from ~4.11 M ha to ~4.53 M ha between 2000 and 2013 with more prominent increases in African nations such as Burundi, Cameroon, and Ghana (Table 1). In Central and South America, oil palm cultivated area increased by 118% from ~435,000 ha to ~989,000 ha between 2000 and 2013 (Table 2) (FAOSTAT 2015).

Countries with more than twice an increase in oil palm cultivated area include Mexico, Peru, Honduras, Guatemala, Nicaragua and Brazil (Table 2). The way monoculture oil palm agriculture expands in these tropical developing nations will determine whether a legacy of positive or negative socio-economic and the environmental impacts are left behind in these countries (Rival and Levang 2014). Rapid expansion of monoculture oil palm plantations in Southeast Asia has generated a large amount of controversy, which involves the clearing of biodiverse and carbon rich forests and peatlands as well as land grabbing for oil palm plantation development from local communities (Sheil et al., 2009; Rival and Levang 2014). However governments of many developing tropical countries are highly interested in oil palm agriculture because the sector is seen as a unique opportunity to achieve economic and development goals (World Bank 2011; Rival and Levang 2014). Both Indonesia and Malaysia view the oil palm sector as a pillar of economic growth for their national economies (World Bank 2011), while rural employment opportunities and income generation have real and concrete benefits for smallholders (Zen et al., 2008; Rist et al., 2010; Lee et al., 2013b). Regions such as Western and Central Africa, as well as Latin America, have already begun investing into the oil palm sector. In Colombia, for instance, the sector had considered a five-fold increase in total planted area, although recent political changes and the pread of a fungal disease has truncated this plan. Other countries like Liberia have granted oil palm industries large-scale concessions (up to 200,000 ha) to be developed for oil palm plantations (Balachandran et al., 2012). Due to the controversy of rapid oil palm expansion in Southeast Asia, many concerns have been raised over the possible consequences of a new wave of rapid oil palm expansion in tropical Africa and Central and South America (Gutiérrez-Vélez et al., 2011; Sayer et al., 2012; Wich et al., 2014).

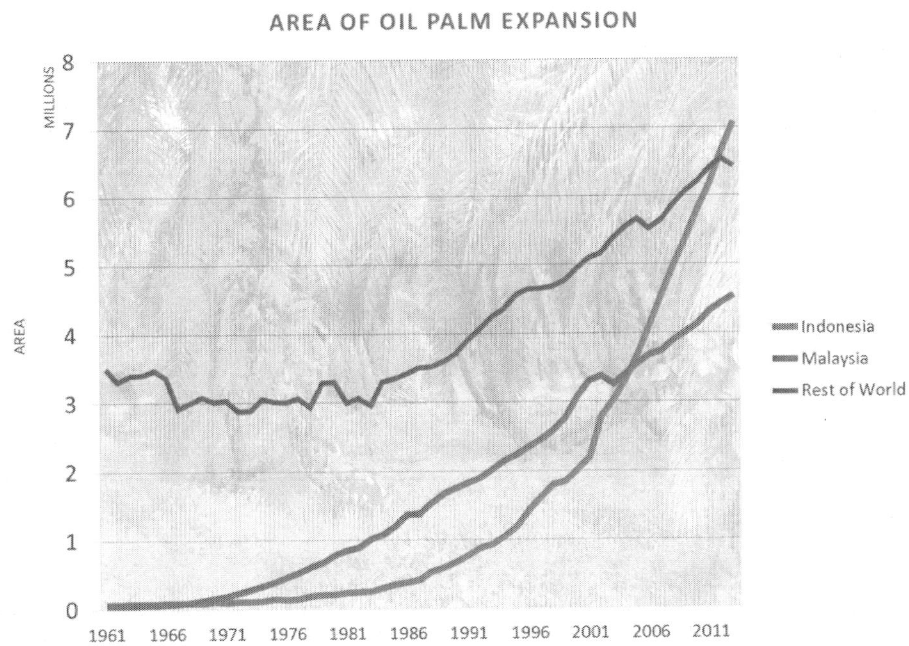

Figure 1. Area of oil palm expansion for Indonesia, Malaysia and Rest of the World from 1961 to 2013.

Table 1. List of African countries with cultivating oil palm

African Countries	Area cultivated in 2000 (ha)	Area cultivated in 2013 (ha)	Percentage Increase in Land Area (%)
Angola	23,000	23,000	0%
Benin	21,000	36,000	71%
Burundi	1,200	7,900	558%
Cameroon	56,075	135,000	141%
Central African Republic	4,000	650	-84%
Congo	7,200	11,600	61%
Côte d'Ivoire	159,000	270,000	70%
Democratic Republic of the Congo	200,000	275,000	38%
Equatorial Guinea	3,500	3,500	0%
Gabon	4,000	4,200	5%
Gambia	3,500	3,500	0%
Ghana	160,000	360,000	125%
Guinea	310,000	310,000	0%
Guinea-Bissau	9,500	9,500	0%
Liberia	16,200	17,000	5%
Madagascar	1,800	1,850	3%
Nigeria	3,080,000	3,000,000	-3%
Sao Tome and Principe	1,100	1,700	55%
Senegal	6,500	11,200	72%
Sierra Leone	21,800	27,500	26%
Togo	13,500	17,000	26%
United Republic of Tanzania	4,500	5,200	16%
Total	4,107,375	4,531,300	10%

Source: FAOSTAT 2015.

Table 2. List of Central and South American countries with cultivating oil palm

Central and South American countries	Area cultivated in 2000 (ha)	Area cultivated in 2013 (ha)	Percentage Increase in Land Area (%)
Brazil	45,000	108,635	141%
Colombia	134,772	250,000	85%
Costa Rica	39,790	74,500	87%
Ecuador	113,686	218,833	92%
Guatemala	21,000	65,000	210%
Honduras	34,000	125,000	268%
Mexico	9,000	46,094	412%
Nicaragua	2,000	5,000	150%
Panama	5,600	5,400	-4%
Paraguay	13,000	16,000	23%
Peru	9,990	47,488	375%
Suriname	550	675	23%
Venezuela (Bolivarian Republic of)	24,711	27,000	9%
Total	453,099	989,625	118%

Source: FAOSTAT 2015.

Here in this chapter we review the environmental impacts of oil palm expansion in developing tropics, taking into account different geographical and management contexts of these expansions. We discuss recent developments in research and in the industry which seek to mitigate the negative environmental impacts of oil palm expansion for multiple stakeholders in the oil palm industry.

ENVIRONMENTAL IMPACTS

Deforestation and Land Use Change

Direct Land Use Change

The expansion of oil palm plantations in the tropics have come at the expense of biodiverse natural habitats which share this landscape. Among these natural habitats, lowland tropical forests have been widely targeted for clearing and planting of oil palm plantations. In Indonesia and Malaysia, the clearing of tropical forests bring an additional benefit by supplying timber through the logging industry (Casson 2000). The revenue brought in from the sale of timber resources have been used to fund the establishment costs of setting up an oil palm plantation (Casson 2000), though this is not always the case. It has been well documented that oil palm plantation development has been used as a front to obtain logging permits in Kalimantan with no subsequent establishment of oil palm plantations thereafter (Casson 2000; Casson and Obidzinski 2002). These impacts on deforestation from oil palm development over the last two decades have been well studied through desktop analyses and satellite imagery analyses.

Using land cover data compiled by the United Nations Food and Agriculture Organization, the extent of forest cover loss as a result of oil palm expansion in Malaysia and Indonesia between 1990 and 2005, Koh and Wilcove (2008) report that ~55%-59% of Malaysia's oil palm expansion occurred at the expense of forests, while in Indonesia at least ~56% occurred at the expense of these ecosystems. In a separate study, Abood et al., (2014) examined forest loss within different type of concession areas in Indonesia, and found that between 2000 to 2010 the oil palm industry was responsible for the largest forest losses in Kalimantan (22%, ~1.1 Mha) and the second largest forest losses in Sumatra (9%, ~440,000 ha), when compared to three other industries (pulp and paper, logging, and mining).

More detailed remote sensing studies also showed rapid conversion of lowland tropical forests to oil palm plantations. Using satellite imagery, it was demonstrated that 90% of lands converted to oil palm from 1990 to 2010 in Kalimantan were forested (~2.84 Mha), with a mixture of intact forests (47%), logged forests (22%), and agroforests (21%) (Carlson et al., 2013). In another remote sensing study conducted by a working group under the Rountable of Sustainable Palm Oil (RSPO), which is a multistakeholder organization consisting of businesses, non-governmental organizations and government bodies, approximately 3.5 Mha of forest in Indonesia, Malaysia and Papua New Guinea was converted for oil palm between 1990 and 2010 (Gunarso et al., 2013). Forest conversion to establish oil palm in this study included both undisturbed and disturbed forest in both upland and swamp forest habitats. The area of forest to oil palm conversion summed over all temporal periods was proportionally greatest in Papua (61% or 33,600 ha), Sabah (62% or 714,000 ha), and Papua New Guinea

(54% or 41,700 ha), followed by Kalimantan (44% or 1.23 Mha), Sarawak (48% or 471,000 ha), Sumatra (25% or 883,000 ha), and Peninsular Malaysia (28% or 318,000 ha) (Gunarso et al., 2013).

Peatlands which are important ecosystems for carbon storage have also been rapidly converted to oil palm plantations (Page et al., 2002; Koh et al., 2011; Carlson et al., 2013). Peatlands are found on low-lying areas close to the coast, making them attractive in terms of redcued transportation costs of palm oil to shipping ports. Moreover, the reduced availability of lowlands results in peatlands being developed despite the higher establishment costs. In Indonesia, many peat swamp forest areas are also less contested in terms of land ownership rights from local communities and have been delineated by district or provincial authorities for industrial plantation development (Sargeant 2011). The development of peatlands for industrial plantations require a process of drainage, where canals are dug out to lower the water table of the peat ecosystem, exposing peat to the atmosphere (Page et al., 2011a). This results in high carbon dioxide emissions due to peat oxidation as well as emissions from peat fires which occur more frequently as compared to when the peat ecosystem is waterlogged (Page et al., 2011a). Koh et al., (2011) demonstrated that 6% (or ~880,000 ha) of tropical peatlands in the region of Peninsular Malaysia, Borneo, and Sumatra had been converted to oil palm plantations by the early 2000s. Based on the RSPO study, by 2010, oil palm plantations on peatlands constituted 18% (2.4 Mha) of the spatial footprint of palm oil (Gunarso et al., 2013). Just within Kalimantan, peatland conversion to oil palm increased from 3% in the 1990s to 16% in the 2000s (Carlson et al., 2013). Often, the impact on peatswamp ecosystems as a result of conversion to oil palm plantations is irreversible and has strong negative impacts on greenhouse gas emissions (Page et al., 2011b) as well as vegetation fires which contribute signifcantly to atmospheric pollution or haze in Southeast Asia (Gaveau et al., 2014).

While oil palm has been expanding steadily in Africa and Latin and South America, the number of studies looking into land cover change and deforestation are still lagging behind Southeast Asia. Historically, these regions face other deforestation pressures such as small-scale land clearing, degradation from illegal logging, and other agricultural drivers such as soybean and sugarcane expansion, as well as cattle ranching (Nepstad et al., 2006; Phalan et al., 2013; Sassen et al., 2013). Few studies also compare the impact of smallholder versus industrial oil palm plantations in the oil palm industry, mainly due to greater difficulty in identifying smallholder oil palm plantations as compared to industrial plantations (Uryu et al., 2008). In the Peruvian Amazon, (Gutiérrez-Vélez et al., 2011) measured the impact of industrial and smallholder oil palm expansion on forests. Using satellite and field data, they demonstrated that 72% (~15,300 ha) of new plantations expanded into forested areas from 2000 to 2010. In a focus area of the Ucayali region, low-yield smallholder plantations accounted for most expansion overall (80%) but only 30% of their expansion involved forest conversion, contrasting with 75% for high-yield industrial oil palm plantations. Uryu et al., (2008) looked into deforestation rates in Riau from 1982 to 2007 and reported higher deforestation from oil palm private enterprises (28%) than oil palm smallholdings (7%). In Sumatra, Lee et al., (2013a) demonstrated that oil palm-driven deforestation from 2000 to 2010 was led predominantly by private enterprises (88%), followed by smallholdings (10%), and state-owned plantations (~1%). As Africa and Latin and South America are considered the next frontiers for oil palm expansion (Wich et al., 2014; Gilroy et al., 2015), and as other stakeholders such as smallholders become more important in the oil palm industry (World

Bank 2011), it is important for researchers to continue to monitor trends of deforestation in these regions and the relative contribution from different stakeholders within the oil palm industry.

Indirect Land Use Change

The expansion of oil palm might not only result in the direct conversion of land, but also lead to indirect land use changes by displacing other agricultural and forestry activities elsewhere. Thus, the conversion of agricultural land to oil palm can become a local or distant driver of land use change in itself, amplifying at broader scales impacts such as habitat loss, fragmentation, habitat degradation and alteration of ecosystem processes. Indirect land-use changes are typical of situations where commodities are produced in one location but consumed in a different one, and are, then, directly linked to globalization, trade, market shifts or unintended results from land-use policies (Meyfroidt et al., 2013). Not only can they occur when oil palm directly displaces crops to other areas to meet demand, but also when the conversion of agricultural land to oil palm triggers higher crop prices, thus, stimulating land conversion to agriculture elsewhere (Wicke et al., 2012).

Accounting for both direct and indirect land-use changes is increasingly becoming important in sustainability policies of agro-industrial activities (Broch et al., 2013; Immerzeel et al., 2014). However, attributing a causal relationship between these changes is a difficult task because of their market-driven nature and broad-scale dynamics (Ziegler et al., 2012; Meyfroidt et al., 2013). Most of our knowledge on this type of impacts originates from studies on the land use transition dynamics between cattle ranching and soy industries in Brazil. In the case of oil palm, very few studies have documented or addressed indirect land use changes. It is clear, however, that they may occur because of the global nature of palm oil trade and because plantations have been established not only at the expense on natural habitats, but also on ranching, agricultural and agroforestry land (Gutiérrez-Vélez et al., 2011; Carlson et al., 2012a; Castiblanco et al., 2013; Dewi et al., 2013). The conversion of agroforestry and other types of agricultural friendly farming poses an additional risk for biodiversity in existing transformed landscapes. Furthermore, studies analysing future sustainable scenarios of oil-palm expansion have identified low-efficient agricultural and degraded land as some of the areas potentially suitable for conversion (Garcia-Ulloa et al., 2012, Koh et al., 2010). These findings, thus, emphasize the importance of appropriate planned expansion in both natural habitats and human-modified land.

Biodiversity

There are various aspects of tropical biodiversity that are impacted as a result from the expansion of oil palm agriculture. These include changes in species composition, richness, abundance and distribution, as well as, in ecological dynamics (e.g., species interactions) or population dyamics (e.g., genetic structure). Changes to these various aspects can ultimately lead to local and global extinction of species, subspecies and populations. The mechanisms in which oil palm expansion influences biodiversity can be direct when the establishment of plantations directly causes changes in biodiversity levels, for instance, by modifying habitat availability in the short and long term. They can also be indirect when these mechanisms operate more diffusely and in no apparent connection to the establishment of a plantation (e.g., when oil palm expansion on agricultural land displaces other crops to natural areas). Biodiversity changes in oil palm landscapes and regions can also be influenced by external factors that are not under the control of growers, communities or decision makers such as climate change and invasive species. The effects of this range of mechanisms are synergistic, cumulative, and contribute to biodiversity changes from local to global scales.

Habitat Loss

Biodiversity impacts from large-scale habitat loss is perhaps the most visible and most frequently cited impact from oil-palm expansion. Land conversion to accommodate plantations has resulted in the loss of vast areas of tropical forests in countries like Indonesia and Malaysia, where in the islands of Borneo and Sumatra approximately 50–85% of industrial oil palm plantations were established on primary or logged forests since 1990 (Miettinen et al., 2012; Carlson et al., 2013). Because many of these ecosystems hold some of the highest levels of biodiversity on the planet, and because oil palm landscapes can only harbour a significantly smaller proportion of this diversity (Fitzherbert et al., 2008), large-scale land conversion to accommodate oil palm can result in the endangerment or extinction of a vast number of species by drastically reducing their habitat. For instance, the range area of all 6 species of great apes overlap greatly with suitable land for oil palm in Africa and Asia (Wich et al., 2012; Wich et al., 2014). Three of these species the Chimpanzee *Pan troglodytes*, Orang-utan *Pongo pygmaeus,* and Bonobo *Pan paniscus*, are regarded as Endangered, while the Gorilla *Gorilla gorilla* is considered Critically Endangered by the IUCN on the grounds of habitat loss among other reasons. Thus, further extensive habitat degradation, driven by oil palm, could potentially worsen the conservation status of these apes.

Habitat Fragmentation and Landscape Connectivity

Large scale conversion of natural habitats for biodiversity not only result in direct habitat losses but also increases habitat fragmentation, creating an irregular network of forests fragments embedded in a human-dominated matrix such as oil palm (Laurance 2014). Fragmentation has been recognized as a driver of biodiversity change at multiple scales (Haddad et al., 2015), and has been an important subject of study in landscape and population ecology (Fahrig 2013). It can drive biodiversity changes by modulating ecological, behavioural and evolutionary processes, for instance, by restricting species movement or decreasing genetic variability of populations (Tscharntke et al., 2012). Thus, these changes are the result of the combined effects of habitat spatial-arrangement in the landscape and ecological aspects of species (Döbert et al., 2014). These effects are cumulative at larger scales, can act synergistically with other drivers and can affect not only biodiversity but also ecosystems processes in the landscape and the regional scale (Haddad et al., 2015).

How spatial configurations of oil palm landscapes influence biodiversity is less well understood as compared to understanding biodiversity patterns within the oil palm plantation (i.e., oil-palm as a habitat) and remnants of natural habitats within the oil palm landscape (Savilaakso et al., 2014). Overall, studies show that the occurrence of natural habitats, together with habitat heterogeneity and structural complexity within the oil palm plantation, can increase biodiversity levels in the oil-palm landscape (Luskin and Potts 2011; Livingston et al., 2013; Gray et al., 2014; Azhar et al., 2014). Other evidence, indicate that biodiversity levels within fragments are not as high as those observed in natural continuous forests, which together with the low biodiversity levels found in the oil palm matrix, suggest that these landscapes as a whole cannot sustain the high biodiversity levels found in natural forests (Edwards et al., 2010). However in a landscape such as the Llanos of Colombia, which is a mixed grassland-forest system, oil palm plantations have a similar or higher species richness across four taxonomic groups (ants, dung beetles, birds and herpetofauna) than improved pasture, suggesting that existing cattle pastures could offer an opportunity to meet governmental targets for oil palm development without incurring significant biodiversity costs (Gilroy et al., 2015). Preserving forest remnants in such landscapes are also valued to protect biodiversity and export avian 'spill-over' effects into oil palm plantations (Gilroy et al., 2015). The connectivity of oil-palm landscapes is an important aspect that has also been overlooked in these studies. It is assumed that landscapes which retain high levels of connectivity are more likely to maintain populations of species that inhabit the original continuous landscape (Haddad et al., 2015). While it has been argued that oil palm plantations can be a permeable land use, the dispersal ecology of forests species within oil palm landscapes is important to facilitate conservation planning. In Borneo, studies have shown that securing and facilitating such dispersals in oil-palm regions should be a key conservation priority for Orangutan and other species (Campbell-Smith et al., 2011; Proctor et al., 2011).

The suitability of oil palm plantations as a habitat is not the only aspect that influences biodiversity persistence in the short and long term. Other spatial aspects, such as quality of the fragmented natural habitat (e.g., size and shape), the effects of boundaries between oil palm plantation and natural habitat fragments (i.e., edge effects) and the connectivity of natural habitat fragments in the landscape, also play important roles in steering biodiversity changes. These spatial aspects are also specific to each taxa, and vary according to species behaviour and ecology. Species that in theory may survive in oil palm landscapes, in reality may not be able to do so because living in these environments may reduce their fitness (e.g., by reducing reproduction rates, increasing vulnerability to competition or predation) (Battin 2004; Chazdon 2014); or because of the increasing human pressures that result in fragmented landscapes (e.g., habitat degradation, wildlife conflicts, hunting and poaching) (Azhar et al., 2013). For instance, although Sumatran tigers could benefit from the hyper-abundance of banded pigs in oil palm landscapes, they rarely persist in these areas as they actively avoid human-landscapes (Imron et al., 2011).

Habitat transformation, the opening of new roads and a population increase in oil-palm landscapes can also lead to an increase in hunting, poaching and wildlife conflicts. In Indonesia, for example, the over-abundance of wild boars around plantations has stimulated the emergence of commercial hunting of this species, which in some cases has resulted in additional pressures to various other species because of unselective hunting techniques (Luskin et al., 2014). Other issues such as poaching, road kills and killings by feral dogs have also been reported as an increasing threat in oil-palm landscapes in Malaysia (Azhar et al., 2013). The growing encroachment of human activities in natural areas, not only increases biodiversity exploitation, but also human-wildlife conflicts. For instance, a recent study documented the case of an isolated population of orang-utans in a mosaic landscape dominated by oil palm in Sumatra, which often raided crops because their remaining habitat did not offer enough resources (Campbell-Smith et al., 2011). In another case, the use of electric fences to protect plantations against elephants incursions in Sabah, has restricted the movement of this species in the landscape reducing the habitat available during flood events (Estes et al., 2012). The authors of the study argued that this not only increases the vulnerability of this elephant population, but it also increases the risk of human-elephant conflict on agricultural land outside the oil-palm plantations.

Biodiversity Enhancements and Ecosystem Functioning within Oil Palm Landscapes

Recent articles try to understand the potential of enhancing biodiversity through structural vegetative changes within the oil palm plantation or oil palm landscape. Some of these enhancements are made explicit in the RSPO's Principles and Criteria (http://www.rspo.org/certification/how-rspo-certification-works) such as retaining High Conservation Value (HCV) habitats or forest fragments within the oil palm plantation, applying integrated pest management and reducing the use of pesticides in the system, as well as protecting water courses and wetlands, including maintaining and restoring appropriate riparian and other buffer zones along water ways. The presence of biodiversity as a result of these vegetation enhancements or improved management practices may maintain certain ecosystem functions or species interactions within the oil palm landscape although this field is subject to much recent active research (Gray et al., 2014; Kerfahi et al., 2014; Edwards et al., 2014a; Gray and Lewis 2014; Gray et al., 2015; McGuire et al., 2015; Fayle et al., 2015).

Koh (2008) study on how insectivorous birds from neighboring patches of forests provide pest control and reduced herbivory rates in oil palm plantations provided a first look into the types of ecosystem functions provided by biodiversity in the oil palm landscape. Subsequently, various studies have looked into how community assemblages may alter in oil palm plantations for invertebrate (Fayle et al., 2010; Gray et al., 2014; Edwards et al., 2014a; Barnes et al., 2014; Gray et al., 2015), fungal (Kerfahi et al., 2014; McGuire et al., 2015), fish (Giam et al., 2015), and amphibian (Faruk et al., 2013; Konopik et al., 2014) communities and what these changes in community composition mean in terms of the ecosystem functions provided by these communities as well as species interactions such as prey-predator dynamics. Most studies point towards a reduction in taxonomic and functional diversity indicating a reduction in ecosystem functioning capacity by these communities as a result of habitat changes from intact or logged forests to oil palm plantations (Kerfahi et al., 2014; Edwards et al., 2014a; Barnes et al., 2014; McGuire et al., 2015). However, reduced diversity was not always followed by a reduced capacity for ecosystem functioning such as dung removal among dung beetle communities in riparian buffer zones (Gray et al., 2014). Alterations of community composition in oil palm plantations also seem to maintain predator-prey interactions in which case new prey sources were exploited by predators (Konopik et al., 2014).

Enhancing the biodiversity value of the oil palm plantation by including forest fragments or riparian buffers within the oil palm landscape can sometimes come at an opportunity cost in terms of oil palm productivity (e.g., area available for oil palm cultivation or ecosystem dis-services such as agricultural pests). For example, the creation of riparian buffer zones necessitate less area within the given oil palm concession area for planting more oil palm trees. However, other benefits such as flood control can be considered in the long-run for the designing of such oil palm landscapes, reducing the economic costs of damaged oil palm trees (Abram et al., 2014). Moreover, riparian buffers do contribute in a significant way towards improving the taxonomic and functional diversity of aquatic taxa and ant communities (Gray et al., 2015; Giam et al., 2015), serving as refuges for the continuing of ecosystem functions such as scavenging rates (Gray et al., 2015). Riparian zones have also been shown to not increase the density of defoliating insects, reducing the concerns of plantation managers on the preservation value of buffer zones (Gray and Lewis 2014). On a landscape scale, the retention of forests within an oil palm landscape seem to have a neutral effect on oil palm yields and do not seem to have an effect on ecosystem services and dis-services (Edwards et al., 2014b). Epiphytes occurring on the trunks of oil palm trees also do not affect the yields of oil palm production and should remain uncut for its biodiversity value (Prescott et al., 2015). Tree-based enrichment in oil palm plots do come at an opportunity costs for smallholders (more trees, lower oil palm yields) but these costs seem to be marginal in intensively managed oil palm plots (Teuscher et al., 2015).

Ecosystem Services

Along with the loss of natural habitats comes the loss of various ecosystem services which are associated with these natural habitats. Ecosystem services are important benefits for human beings that arise from healthily functioning ecosystems such as production of oxygen, soil genesis and water detoxification (MEA 2005) These benefits are often not considered in the decision-making process of land use spatial planning as they are not as traditionally valued as compared to profits attained from the sale of crude palm oil in the market. However these intangible benefits do matter when living conditions of human beings are affected as the health of ecosystems deteriorates. The widespread conversion of tropical forests over mineral and peatlands to oil palm plantations pose major concerns with regards to the health of atmospheric, aquatic and terrestrial systems, especially within Southeast Asia.

Tropical forests over mineral and peatlands are hugely important in their role for carbon sequestration, where carbon dioxide in the atmosphere is absorbed for photosynthesis. The clearing of these forests results in both the increase in levels of greenhouse gases emitted into the atmosphere, and the reduction of the earth's capacity to sequester more carbon. This contributes to the phenomenon of global warming and has long-term impacts on global climate change. Carlson et al., (2013) in their land use change study of forest to oil palm conversion estimated that, assuming fires were used to prepare peatlands for oil palm, Kalimantan plantations would have contributed 21% of Indonesia's total emissions from land cover change in the mid-1990s and 9% in the mid-2000s. If all awarded leases were developed, oil palm in Kalimantan alone would be responsible for 34% of Indonesia's 2020 land-sourced emissions, excluding the emissions as a result of other indirect land cover changes (e.g., wildfires, logging, displaced smallholders). Policy scenario analyses demonstrated that protection of peat and forests are crucial in maximizing the reduction of greenhouse gas emissions (Carlson et al., 2013). In a separate study, researchers reported a sharp rise in emissions from conversion of peatlands for plantations, rising from 26 Mt CO_2 yr^{-1} in the 1990s to 56 Mt CO_2 yr^{-1} between 20001 and 2005, and 88 Mt CO_2 yr^{-1} between 2006 and 2010, as the lack of availability of lowlands increased the use of peatlands for oil palm development (Gunarso et al., 2013). Researchers from this study however suggested that forest degradation is a larger source of emissions than oil palm production and noted that in Indonesia, the largest source of historical CO_2 emissions from land use change was from forest degradation (40%) (Gunarso et al., 2013). Based on their analysis, greenhouse gas emissions from land use change and peat oxidation between 2006 and 2010 showed the oil palm sector contributed 16% of total emissions from land use and land use change in Indoneisa and 32% in Malaysia (Gunarso et al., 2013). They do suggest that while a relatively smaller area of plantations (15%) were established on peatlands, this had a disproportionate impact on greenhouse gas emissions since the drainage of peat soils releases large amounts of carbon into the atmosphere (Page et al., 2011a). Based on figures from 2010, plantations on peat constituted 18% (2.4 M ha) of the spatial footprint of palm oil, but emission from peat fires and peat oxidation were the source of ~64% (118 Mt CO_2 yr^{-1}) of the total eissions from land use linked to industrial scale oil palm plantations (Agus et al., 2013).

Apart from greenhouse gas emissions, the use of fires to clear land for oil palm establishment has also led to severe air pollution in the regions where fires occur, as well as in regions much further away than the source of fires (Glover and Jessup 1999; Quah 2002; Aiken 2004). Fires are often used as the cheapest method to clear land for large-scale commercial agriculture and small-scale farming during the dry season in Southeast Asia (Glover and Jessup 1999). As the price of palm oil has increased over the last few decades, the clearance has been largely attributed to the agricultural development of oil palm plantations both at the large and small scale (Ekadinata et al., 2013; Gaveau et al., 2014; Marlier et al., 2015). The management practices during the establishment and operations of oil palm plantations can greatly modify the terrestrial system (especially peat ecosystems) on which these plantations were established on. In West Kalimantan, wildfires escaping from oil palm plantations have likely contributed to forest loss during the El Niño-Southern Oscillation (ENSO) years more than the actual establishment of plantations in the area (Carlson et al., 2012b). Extensive drainage of peat ecosystems have reduced the water tables of these systems, leaving dried peat exposed and susceptible to fires (van der Werf et al., 2008; Page et al., 2011a). A recent analysis on the severe haze episode in Southeast Asia revealed that burning of degraded peatlands accounted for more than 80% of the fires in June 2013, when Singapore's Pollution Standard Index reached 400 (values above 300 deemed hazardous), and Malaysia's Air Pollution Index reached 750 (values above 300 deemed hazardous) (Gaveau et al., 2014). Smoke haze, caused by vegetation and peat fires in Southeast Asia, results in the release of hazardous gases and aerosols, as well as particulate matter with dimensions less than 2.5 μm ($PM_{2.5}$) (Reddington et al., 2014). Particulate emissions are associated with adverse health effects and increased mortality (Emmanuel 2000; Kunii et al., 2010; Johnston et al., 2012; Marlier et al., 2015). It is estimated that in each year where large fires occurred, particulate emissions from fires across Southeast Asia caused 10,800 premature deaths (Marlier et al., 2012).

Oil palm cultivation also affect soil properties, including organic content, water holding capacity (Nurulita et al., 2014) as well as changes in bacterial biomass and diversity (Carron et al., 2015; Krashevska et al., 2015). Plantations required large quantities of water that can result in extensive hydrological changes caused by storage, abstraction and diversion of river flows. These alterations are major threats to wetland ecosystems and their effects to terrestrial ecosystems are largely unknown (Gopal 2013). Oil-palm development can drastically alter disturbance regimes. For instance in the Lower Kinabatangan district in Sabah, Abram et al., (2014) demonstrated that up to 54% of areas allocated plantations are unsuitable for oil palm due to inundation events. Poor plantation design around rivers and in flood plains, thus, can result in both negative economic and environmental effects such as loss of palm trees, erosion and sedimentation. Instead of converting flood-prone regions for oil palm development, these areas could be potentially used as ecological corridors and as protection areas against flooding events (Abram et al., 2014). Peatlands which have been drained and developed for oil palm plantations can also lead to land subsidence as a result of peat shrinkage and biological oxidation (Hooijer et al., 2012). This has implications on the hydrological integrity of the peat system, its susceptibility to burning and flooding events (Wösten et al., 2006; Wösten et al., 2008), as well as the impact of land availability in coastal regions (Deltares 2015).

The development of oil palm plantations in the tropics may also affect freshwater quality and availability for local communities (Donald 2004; Carlson et al., 2014). By comparing streams which drained watersheds of five different land uses (intact forest, logged forest, mixed agroforest, young (<3 years) and mature (>10 years) oil palm plantations), Carlson et al., (2014) quantified the impacts of intensifying land use on freshwater systems from 2009 to 2012 in West Kalimantan. They showed that streams flowing out of young oil palm plantations had higher mean sediment concentrations (fourfold to 550-fold) as compared to all other streams. Streams flowing out of young and mature oil palm plantations also had warmer temperatures (3.0-3.9°C) compared to streams flowing from intact forests (25°C). Ecosystem respiration rates in these streams were also influenced by low rainfall, where mature oil palm streams consumed up to 21 g O_2 m^{-2} d^{-1} in an ENSO drought period, in contrast to 2.8 g O_2 m^{-2} d^{-1} during non-drought periods (Carlson et al., 2014). Impact on freshwater quality can also be exacerbated when waste products and by-products of industrial processing of palm oil into crude palm oil is released into water bodies without appropriate treatment. Palm oil mill effluent (POME), which is the waste product from crude palm oil processing, is rich in organic matter and can contribute to eutrophication and depletion of dissolved oxygen levels in aquatic systems if left untreated (Donald 2004). POME also releases methane, which is a greenhouse gas that has a greater global warming potential than carbon dioxide (Taylor et al., 2014). The negative impacts on freshwater availability and accessibility for local communities surrounding oil palm plantations as well as downstream users of a watershed that is heavily used by oil palm agriculture are known anecdotally (Kesaulija et al., 2014) but would benefit from more research.

CONCLUSION

The environmental impacts from oil palm expansion in the developing tropics are wide ranging and are often a result of poor management of natural resources as well as spatial planning of land uses (Sheil et al., 2009; Rival and Levang 2014). Here, we discuss some solutions and initiatives from various stakeholders of the oil palm industry which have been set up to mitigate the negative environmental impacts from oil palm development, as well as to avoid the mistakes of land use planning for oil palm development in Southeast Asia.

Sustainable Certification of Palm Oil Production

As consumers become more aware of the link between negative environmental impacts and the commercial production of palm oil in their products, the oil palm industry have come under increasing global scrutiny for their impact in tropical developing nations (Khor 2011). In response to this intense public scrutiny and consumer pressure to 'green' their supply chains, many companies have adopted sustainable certification policies to ensure their supply chains are rid of products derived from practices that could be harmful to the environment (Laurance et al., 2010). Certification schemes for palm oil include the international multistakeholder initiative, the Roundtable of Sustainable Palm Oil (RSPO), as well as national initiatives such as the Indonesian Sustainable Palm Oil (ISPO), and the Malaysian Sustainable Palm Oil (MSPO) certification standard (Adnan 2015).

Certification schemes typically comprise a set of environmental and/or social criteria that are directed by principles of sustainability as agreed upon by relevant stakeholders (Lee et al., 2011). These principals and criteria are driven by the overall goal of maintaining natural resources and the supply of ecosystem goods and services to meet present needs without compromising the ability of future generations to meet their own needs. Almost all certification schemes include the protection of biodiversity as a criterion to be considered in the production process. Hunting and protection of endangered species are forbidden whilst landscape management of plantations to preserve as much of native vegetation as possible is encouraged under various certification schemes. The fulfillment of certification criteria – usually verified by independent auditors – results in the certification of a product or production process as "sustainable" (van Dam et al., 2010; McBride et al., 2011) which frequently confers a price premium (Chen et al., 2010).

The RSPO is the oldest and leading certification scheme for palm oil among the three and is also the international standard for palm oil sustainable certification. However, the RSPO often addresses to different degrees the majority of impacts identified from the expansion of this crop at the plantation or plot level. This may indeed yield positive benefits such as the preservation of riparian buffers for improved aquatic and invertebrate diversity (Gray et al., 2014; Gray et al., 2015; Giam et al., 2015). Specific standards within the RSPO also discourage hunting which can have a positive influence on hunting behavior within certified oil palm landscapes (Azhar et al., 2013). However, if oil palm landscapes are to be considered as a part of comprehensive consevration strategies, many of the standards for biodiversity best management practices at the plot-level may not take into sufficient consideration the heterogeneity required of the surrounding landscapes of the oil palm plantation (Azhar et al., 2015). For example, the 'biodiversity-friendliness' of certified large-scale oil palm plantations when compared to uncertified smallholdings in Peninsular Malaysia was called into question as these certified plantations exhibited lower levels of landscape heterogeneity (Azhar et al., 2015). It is understandable that such an expansion of spatial scale would be at the expense of time and resources of plantation and sustainability managers. Hence, certification standards under the RSPO have a limited role to tackle the broader impacts of oil-palm agriculture at larger scales. Additional mechanisms that promote an integrated land-use planning approach to oil palm landscapes and producing regions are needed (Koh et al., 2009), if the negative cumulative impacts from these initiatives are to be minimized and the positive benefits maximized.

Intensifying Oil Palm Production

Intensification is increasingly seen as a broad strategy to minimize the direct and indirect impacts of land conversion on natural habitats by reducing the amount of land needed to meet global demand of palm oil (Carrasco et al., 2014). Recent genome sequencing of the oil palm crop has identified the gene responsible for increasing yields, which has great implications for substantial yield improvements (Singh et al., 2013). For example, Sime Darby, which is a leading Malaysian oil palm company, has used molecular breeding to produce "super palms" variety which can increase yields from 4 to 10 tons/ha. Oil palm genome sequencing may also be used to generate seeds that are resistant to diseases, drought, and salinity in order to increase the resilience of the oil palm crop in difficult environments. However, the conservation benefits of intensifying yields for oil palm production may not be so straight forward.

Improving oil palm yields may appear to be beneficial for conservation. The global demand for palm oil could in theory be met on less land by raising yields, hence sparing land for biodiversity conservation (Phalan et al., 2014). Based on estimates on palm oil consumption and global population growth, it is expected that by 2050, the demand for palm oil will rise substantially and moderate improvements to yields (from 4 to 5.2 tons/ha/year) could lead to a global saving of 7 M ha in Indonesia (Corley 2009). However, when factors such as international trade as well as the substituitability of palm oil as a vegetable oil are considered, the consequences for conservation are varied (Villoria et al., 2013). While yield improvements will leave some ~400,000 ha of land in Brazil, India, and Canada untouched, it would spur an expansion of 65,000 ha for oil palm and 47,000 ha for pastures at the expense of forests in Malaysia and Indonesia (Villoria et al., 2013) as a result of economic benefits from these regions. Moreoever, increased yields may increase the agricultural rent of forested lands, spurring the expansion of oil palm agriculture under such scenarios (Phelps et al., 2013). The intensification of oil palm plantations can be achieved by improved practices such as water irrigation, adequate fertilizer use, better disease and pest control, the use of high-yielding palm varieties or by converting mixed-agricultural or natural-agriculture mosaic areas into strictly monocultures. Intensification might, thus, generate further environmental degradation through higher pollution levels, exhaustive use of soils and water, further homogenization of the vegetation in the landscape, or by indirectly driving additional expansion of this crop because of potentially higher profits (Fischer et al., 2008; Fischer et al., 2011; Phelps et al., 2013; Carrasco et al., 2014).

Spatial Planning and Expansion over Degraded Lands

Oil palm cultivation has expanded in a wide range of forest habitats (including peatland, mangrove and riparian forests), with different conservation or successional states (e.g., primary, secondary, degraded forests) (Carlson et al., 2012a). Much of the conservation debate has, thus, focused on defining the types of habitats where conversion should be avoided and identifying land-uses where oil palm development should be directed. Habitats such as primary and peatland forests have been at the centre of conservation efforts because of their crucial role in supporting biodiversity (Gibson et al., 2011), and storing carbon stocks (Gibbs et al., 2010; Carlson and Curran 2013). The situation is less clear for other habitat types such as degraded and secondary forests, because of the difficulties in defining these habitats (Ghazoul et al., 2015), and because many argue that focusing on the conservation or rehabilitation of non-pristine areas may undermine efforts for the conservation of critically important habitats by spreading resources too thin (Gibson et al., 2011). Secondary forest are, however, increasingly being recognized as important landscape features that could lower and buffer the impacts of habitat loss and fragmentation (Edwards et al., 2011), but their ecological role within oil palm landscapes still needs further research (Savilaakso et al., 2014). An oil palm suitability tool developed by the World Resources Institute enable companies to identify which lands are less rich in carbon and biodiversity levels (http://www.wri.org/resources/maps/suitability-mapper). Such tools inform decision-makers of which areas would be most suitable for the production of oil palm in a sustainable manner (Smit et al., 2013). Other modeling tools perform scenario analyses to understand what the environmental and socioeconomic tradeoffs associated with different spatial land use planning of oil palm expansion in regions (Wicke et al., 2011), countries (Koh and Ghazoul 2010; Garcia-Ulloa et al., 2012), as well in more specific areas (Carlson et al., 2012a; Smit et al., 2013; Lee et al., 2014).

In Southeast Asia the oil palm sector is also approaching a new phase where the first generations of commercial plantations have passed peak yields and are due to be replanted (Snaddon et al., 2013). Many of these plantations were established in environmentally sensitive areas, e.g., peatlands, steep slopes or riparian areas, and their design do not comply with current environmental standards for the sector. Replanting in these areas should be, then, avoided because of the long term effects such as continuous emissions and soil erosion. From an ecological perspective, adequate replanting planning is an opportunity to improve habitat availability and configuration in oil-palm landscapes, through the reforestation of corridors and buffer areas (Koh et al., 2009). Furthermore, many established oil-palm landscapes have already developed more complex species communities and ecological relations. Replanting in such cases should consider potential consequences to those communities (Snaddon et al., 2013).

References

Abood S a., Lee JSH, Burivalova Z, et al., (2014) Relative contributions of the logging, fiber, oil palm, and mining industries to forest loss in Indonesia. Conserv Lett n/a–n/a. doi: 10.1111/conl.12103.

Abram NK, Xofis P, Tzanopoulos J, et al., (2014) Synergies for Improving Oil Palm Production and Forest Conservation in Floodplain Landscapes. *PLoS One* 9:e95388.

Adnan H (2015) Newly-implemented Malaysian Sustainable Palm Oil means more business. Jakarta Post.

Agus F, Gunarso P, Sahardjo BH, et al., (2013) Historical CO2 Emissions from Land Use and Land Use Change from the Oil Palm Industry in Indonesia, Malaysia and Papua New Guinea.

Aiken RS (2004) Runaway Fires, Smoke-Haze Pollution, and Unnatural Disasters in Indonesia. *Geogr Rev* 94:55–79.

Azhar B, Lindenmayer DB, Wood J, et al., (2014) Ecological impacts of oil palm agriculture on forest mammals in plantation estates and smallholdings. *Biodivers Conserv* 23:1175–1191. doi: 10.1007/s10531-014-0656-z.

Azhar B, Lindenmayer DB, Wood J, et al., (2013) Contribution of illegal hunting, culling of pest species, road accidents and feral dogs to biodiversity loss in established oil-palm landscapes. *Wildl Res* 40:1–9. doi: 10.1071/wr12036.

Azhar B, Saadun N, Puan CL, et al., (2015) Promoting landscape heterogeneity to improve the biodiversity benefits of certified palm oil production: Evidence from Peninsular Malaysia. *Glob Ecol Conserv* 3:553–561. doi: 10.1016/j.gecco.2015.02.009.

Balachandran L, Herb E, Timirzi S, O'Reilly E (2012) Everyone must eat? Liberia, Food Security and Palm Oil. In: *Global Land Grabbing II. Ithaca,* NY.

Barnes AD, Jochum M, Mumme S, et al., (2014) Consequences of tropical land use for multitrophic biodiversity and ecosystem functioning. *Nat Commun* 5:5351. doi: 10.1038/ncomms6351.

Battin J (2004) When Good Animals Love Bad Habitats: Ecological Traps and the Conservation of Animal Populations. *Conserv Biol* 18:1482–1491. doi: 10.1111/j.1523-1739.2004.00417.x.

Broch A, Hoekman SK, Unnasch S (2013) A review of variability in indirect land use change assessment and modeling in biofuel policy. *Environ Sci Policy* 29:147–157. doi: 10.1016/j.envsci.2013.02.002.

Campbell-Smith G, Campbell-Smith M, Singleton I, Linkie M (2011) Apes in space: Saving an imperilled orangutan population in Sumatra. *PLoS One.* doi: 10.1371/journal.pone.0017210.

Carlson KM, Curran LM (2013) Refined carbon accounting for oil palm agriculture: disentangling potential contributions of indirect emissions and smallholder farmers. *Carbon Manag* 4:347–349. doi: 10.4155/cmt.13.39.

Carlson KM, Curran LM, Asner GP, et al., (2013) Carbon emissions from forest conversion by Kalimantan oil palm plantations. *Nat Clim Chang* 3:283–287. doi: DOI:10.1038/nclimate1702.

Carlson KM, Curran LM, Ponette-González AG, et al., (2014) Influence of watershed-climate interactions on stream temperature, sediment yield, and metabolism along a land use intensity gradient in Indonesian Borneo. *J Geophys Res Biogeosciences* 119:1110–1128. doi: 10.1002/2013JG00 2516.

Carlson KM, Curran LM, Ratnasari D, et al., (2012a) Committed carbon emissions, deforestation, and community land conversion from oil palm plantation expansion in West Kalimantan, Indonesia. *Proc Natl Acad Sci* 109:7559–7564. doi: 10.1073/pnas.1200452109.

Carlson KM, Curran LM, Ratnasari D, et al., (2012b) Committed carbon emissions, deforestation, and community land conversion from oil palm plantation expansion in West Kalimantan, Indonesia. *Proc Natl Acad Sci U S A* 109:7559–7564. doi: DOI: 10.1073/pnas.1200452109.

Carrasco LR, Larrosa C, Milner-Gulland EJ, Edwards DP (2014) Conservation. A double-edged sword for tropical forests. *Science* 346:38–40. doi: 10.1126/science.1256685.

Carron MP, Auriac Q, Snoeck D, et al., (2015) Spatial heterogeneity of soil quality around mature oil palms receiving mineral fertilization. *Eur J Soil Biol* 66:24–31. doi: 10.1016/j.ejsobi.2014.11.005.

Casson A (2000) The Hesitant Boom: Indonesia's Oil Palm Sub-Sector in an Era of Economic Crisis and Political Change. Center for International Forestry Research, Bogor, Indonesia.

Casson A, Obidzinski K (2002) From New Order to Regional Autonomy: Shifting Dynamics of "Illegal" Logging in Kalimantan, Indonesia. *World Dev* 30:2133–2151. doi: http://dx.doi.org/10.1016/S0305-750X(02)00125-0.

Castiblanco C, Etter A, Aide TM (2013) Oil palm plantations in Colombia: a model of future expansion. *Environ Sci Po*licy 27:172–183. doi: 10.1016/j.envsci.2013.01.003.

Chazdon RL (2014) Second growth. The promise of tropical forest regeneration in an age of deforestation. The University of Chicago Press., Chicago, IL.

Chen J, Innes JL, Tikina A (2010) Private Cost-Benefits of Voluntary Forest Product Certification. *Int For Rev* 12:1–12. doi: 10.1505/ifor.12.1.1.

Corley RH V (2009) How much palm oil do we need? *Environ Sci Policy* 12:134–139. doi: 10.1016/j.envsci.2008.10.011.

Corley RH V, Tinker PBH (2003) The Oil Palm (World Agriculture Series). Balckwell Publishing Limited, Oxford.

Deltares (2015) Flooding projections from elevation and subsidence models for oil palm plantations in Rajang Delta peatlands, Sarawak, Malaysia.

Dewi S, van Noordwijk M, Ekadinata A, Pfund J-L (2013) Protected areas within multifunctional landscapes: Squeezing out intermediate land use intensities in the tropics? *Land use policy* 30:38–56. doi: http://dx.doi.org/ 10.1016/j.landusepol.2012.02.006.

Döbert TF, Webber BL, Barnes AD, et al., (2014) Forest fragmentation and biodiversity conservation in human-dominated landscapes. In: Kettle CJ, Koh LP (eds) *Global Forest Fragmentation*. CABI.

Donald PF (2004) Biodiversity impacts of some agricultural commodity production systems. *Conserv Biol* 18:17–37.

Edwards DP, Hodgson JA, Hamer KC, et al., (2010) Wildlife-friendly oil palm plantations fail to protect biodiversity effectively. *Conserv Lett* 3:236–242. doi: 10.1111/j.1755-263X.2010.00107.x.

Edwards DP, Larsen TH, Docherty TDS, et al., (2011) Degraded lands worth protecting: the biological importance of Southeast Asia's repeatedly logged forests. *P Roy Soc B-Biol Sci* 278:82–90. doi: 10.1098/rspb.2010.1062.

Edwards FA, Edwards DP, Larsen TH, et al., (2014a) Does logging and forest conversion to oil palm agriculture alter functional diversity in a biodiversity hotspot? *Anim Conserv* 17:163–173. doi: 10.1111/acv.12074.

Edwards FA, Edwards DP, Sloan S, Hamer KC (2014b) Sustainable management in crop monocultures: the impact of retaining forest on oil palm yield. *PLoS One* 9:e91695. doi: 10.1371/journal.pone.0091695.

Ekadinata A, van Noordwijk M, Budidarsono S, Dewi S (2013) Hot spots in Riau, haze in Singapore: the June 2013 event analyzed. ASB Partnership for the Tropical Forest Margins. World Agroforestry Centre (ICRAF), Nairobi, Kenya.

Emmanuel SC (2000) Impact to lung health of haze from forest fires: The Singapore experience. *Respirology* 5:175–182. doi: 10.1046/j.1440-1843.2000.00247.x.

Estes JG, Othman N, Ismail S, et al., (2012) Quantity and Configuration of Available Elephant Habitat and Related Conservation Concerns in the Lower Kinabatangan Floodplain of Sabah, Malaysia. *PLoS One.* doi: 10.1371/journal.pone.0044601.

Fahrig L (2013) Rethinking patch size and isolation effects: the habitat amount hypothesis. *J Biogeogr* 40:1649–1663.

FAO (2015) World Agriculture: towards 2015/2030. Rome.

FAOSTAT (2013) FAOSTAT Online Statistical Service. http://faostat.fao.org/.

FAOSTAT (2015) FAOSTAT Production data. In: *FAO Stat. Div.* http://faostat3.fao.org.

Faruk A, Belabut D, Ahmad N, et al., (2013) Effects of oil-palm plantations on diversity of tropical anurans. *Conserv Biol* 27:615–24. doi: 10.1111/cobi.12062.

Fayle TM, Edwards DP, Foster WA, et al., (2015) An ant-plant by-product mutualism is robust to selective logging of rain forest and conversion to oil palm plantation. *Oecologia* 178:441–50. doi: 10.1007/s00442-014-3208-z.

Fayle TM, Turner EC, Snaddon JL, et al., (2010) Oil palm expansion into rain forest greatly reduces ant biodiversity in canopy, epiphytes and leaf-litter. *Basic Appl Ecol* 11:337–345. doi: 10.1016/j.baae.2009.12.009.

Fischer J, Batáry P, Bawa KS, et al., (2011) Conservation: Limits of Land Sparing. *Science* (80-) 334:593. doi: 10.1126/science.334.6056.593-a.

Fischer J, Brosi B, DAILY GC, et al., (2008) Should agricultural policies encourage land sparing or wildlife-friendly farming? *Front Ecol Environ* 6:380–385. doi: 10.1890/070019.

Fitzherbert EB, Struebig MJ, Morel A, et al., (2008) How will oil palm expansion affect biodiversity? *Trends Ecol Evol* 23:538–545.

Garcia-Ulloa J, Sloan S, Pacheco P, et al., (2012) Lowering environmental costs of oil-palm expansion in Colombia. *Conserv Lett* 5:366–375. doi: 10.1111/j.1755-263X.2012.00254.x.

Gaveau DLA, Salim MA, Hergoualc'h K, et al., (2014) Major atmospheric emissions from peat fires in Southeast Asia during non-drought years: evidence from the 2013 Sumatran fires. *Sci Rep* 4:6112. doi: 10.1038/srep06112.

Ghazoul J, Burivalova Z, Garcia-Ulloa J, King L (2015) Conceptualising forest degradation.

Giam X, Hadiaty RK, Tan HH, et al., (2015) Mitigating the impact of oil-palm monoculture on freshwater fishes in Southeast Asia. *Conserv Biol. doi:* 10.1111/cobi.12483.

Gibbs HK, Ruesch AS, Achard F, et al., (2010) Tropical forests were the primary sources of new agricultural land in the 1980s and 1990s. *Proc Natl Acad Sci U S A* 107:16732–16737. doi: DOI: 10.1073/pnas.0910275107.

Gibson L, Lee TM, Koh LP, et al., (2011) Primary forests are irreplaceable for sustaining tropical biodiversity. *Nature* 478:378–381.

Gilroy JJ, Prescott GW, Cardenas JS, et al., (2015) Minimizing the biodiversity impact of Neotropical oil palm development. *Glob Chang Bio*l 21:1531–40. doi: 10.1111/gcb.12696.

Glover D, Jessup T (1999) Indonesia's Fires and Haze: The Cost of Catastrophe. International Development Research Center and the Institute of Southeast Asian Studies, Ottawa and Singapore.

Gopal B (2013) Future of wetlands in tropical and subtropical Asia, especially in the face of climate change. *Aquat Sci* 75:39–61. doi: 10.1007/s00027-011-0247-y.

Gray CL, Lewis OT (2014) Do riparian forest fragments provide ecosystem services or disservices in surrounding oil palm plantations? *Basic Appl Ecol* 15:693–700. doi: 10.1016/j.baae.2014.09.009.

Gray CL, Lewis OT, Chung AYC, Fayle TM (2015) Riparian reserves within oil palm plantations conserve logged forest leaf litter ant communities and maintain associated scavenging rates. *J Appl Ecol* 52:31–40. doi: 10.1111/1365-2664.12371.

Gray CL, Slade EM, Mann DJ, Lewis OT (2014) Do riparian reserves support dung beetle biodiversity and ecosystem services in oil palm- dominated tropical landscapes? *Ecol Evol* 4:1049–1060. doi: 10.1002/ece3.1003.

Guardian Sustainable Business (2014) From rainforest to your cupboard: the real story of palm oil - interactive. In: Guard. http://www.theguardian. com/sustainable-business/ng-interactive/2014/nov/10/palm-oil-rainforest-cupboard-interactive. Accessed 1 Jul 2015.

Gunarso P, Hartoyo ME, Agus F, Killeen TJ (2013) Oil palm and land use change in Indonesia, Malaysia and Papua New Guinea.

Gutiérrez-Vélez VH, DeFries R, Pinedo-Vásquez M, et al., (2011) High-yield oil palm expansion spares land at the expense of forests in the Peruvian Amazon. *Environ Res Lett* 6:44029. doi: DOI: 10.1088/1748-9326/6/4/ 044029.

Haddad NM, Brudvig L a, Clobert J, et al., (2015) Habitat fragmentation and its lasting impact on Earth ecosystems. *Sci Adv* 1–9. doi: 10.1126/ sciadv.1500052.

Hooijer A, Page S, Jauhiainen J, et al., (2012) Subsidence and carbon loss in drained tropical peatlands. *Biogeosciences* 9:1053–1071.

Immerzeel DJ, Verweij P, van der Hilst F, Faaij APC (2014) Biodiversity impacts of bioenergy crop production: a state-of-the-art review. *Glob Chang Biol Bioenergy* 6:183–209. doi: 10.1111/gcbb.12067.

Imron MA, Herzog S, Berger U (2011) The influence of agroforestry and other land-use types on the persistence of a Sumatran tiger (Panthera tigris sumatrae) population: an individual-based model approach. *Environ Manage* 48:276–88. doi: 10.1007/s00267-010-9577-0.

Johnston FH, Henderson SB, Chen Y, et al., (2012) Estimated Global Mortality Attributable to Smoke from Landscape Fires. *Environ Health Perspect* 120:695–701. doi: 10.2307/41548667.

Kerfahi D, Tripathi BM, Lee J, et al., (2014) The impact of selective-logging and forest clearance for oil palm on fungal communities in Borneo. *PLoS One* 9:e111525. doi: 10.1371/journal.pone.0111525.

Kesaulija FF, Sadsoeitoebeon BMG, Peday HFZ, et al., (2014) Oil palm estate development and its impact on forests and local communities in West Papua. A case study on the Prafi Plain.

Khor YK (2011) The oil palm industry bows to NGO campaigns. *Lipid Technol* 23:102–104.

Koh LP (2008) Birds defend oil palms from herbivorous insects. *Ecol Appl* 18:821–825.

Koh LP, Ghazoul J (2010) Spatially explicit scenario analysis for reconciling agricultural expansion, forest protection, and carbon conservation in Indonesia. *Proc Natl Acad Sci* 107:11140–11144. doi: 10.1073/pnas. 1000530107.

Koh LP, Lee TM (2012) Sensible consumerism for environmental sustainability. Biol Conserv 151:3–6. doi: 10.1016/j.biocon.2011.10.029.

Koh LP, Levang P, Ghazoul J (2009) Designer landscapes for sustainable biofuels. *Trends Ecol Evol* 24:431–438. doi: DOI 10.1016/j.tree. 2009.03.012.

Koh LP, Miettinen J, Liew SC, Ghazoul J (2011) Remotely sensed evidence of tropical peatland conversion to oil palm. *Proc Natl Acad Sci U S A* 108:5127–5132.

Koh LP, Wilcove DS (2008) Is oil palm agriculture really destroying tropical biodiversity? *Conserv Lett* 1:60–64.

Konopik O, Gray CL, Grafe TU, et al., (2014) From rainforest to oil palm plantations: Shifts in predator population and prey communities, but resistant interactions. *Glob Ecol Conserv* 2:385–394. doi: 10.1016/ j.gecco.2014.10.011.

Krashevska V, Klarner B, Widyastuti R, et al., (2015) Impact of tropical lowland rainforest conversion into rubber and oil palm plantations on soil microbial communities. *Biol Fertil Soils.* doi: 10.1007/s00374-015-1021-4.

Kunii O, Kanagawa S, Yajima I, et al., (2010) The 1997 haze disaster in Indonesia: its air quality and health effects. *Arch Environ Health* 57:16–22. doi: 10.1080/0003989 0209602912.

Laurance WF (2014) Contemporary drivers of habitat fragmentation. In: Kettle CJ, Koh LP (eds) *Global Forest Fragmentation.* CABI International, Wallingford.

Laurance WF, Koh LP, Butler R, et al., (2010) Improving the Performance of the Roundtable on Sustainable Palm Oil for Nature Conservation. *Conserv Biol* 24:377–381. doi: 10.1111/j.1523-1739.2010.01448.x.

Lee JSH, Abood S, Ghazoul J, et al., (2013a) Environmental impacts of large-scale oil palm enterprises exceed that of smallholdings in Indonesia. *Conserv Lett* doi: 10.1111/conl.12039. doi: 10.1111/conl.12039.

Lee JSH, Garcia-Ulloa J, Ghazoul J, et al., (2014) Modelling environmental and socio-economic trade-offs associated with land-sparing and land-sharing approaches to oil palm expansion. *J Appl Ecol* 51:1366–1377. doi: 10.1111/1365-2664.12286.

Lee JSH, Ghazoul J, Obidzinski K, Koh LP (2013b) Oil palm smallholder yields and incomes constrained by harvesting practices and type of smallholder management in Indonesia. *Agron Sustain Dev.* doi: 10.1007/s13593-013-0159-4.

Lee JSH, Rist L, Obidzinski K, et al., (2011) No farmer left behind in sustainable biofuel production. *Biol Conserv* 144:2512–2516.

Livingston G, Jha S, Vega A, Gilbert L (2013) Conservation Value and Permeability of Neotropical Oil Palm Landscapes for Orchid Bees. *PLoS One* 8:e78523. doi: 10.1371/journal.pone.0078523.

Luskin MS, Christina ED, Kelley LC, Potts MD (2014) Modern Hunting Practices and Wild Meat Trade in the Oil Palm Plantation-Dominated Landscapes of Sumatra, Indonesia. *Hum Ecol* 42:35–45. doi: 10.1007/s10745-013-9606-8.

Luskin MS, Potts MD (2011) Microclimate and habitat heterogeneity through the oil palm lifecycle. *Basic Appl Ecol* 12:540–551. doi: 10.1016/ j.baae.2011.06.004.

Marlier ME, DeFries RS, Kim PS, et al., (2015) Regional air quality impacts of future fire emissions in Sumatra and Kalimantan. *Environ Res Lett* 10:054010. doi: 10.1088/1748-9326/10/5/054010.

Marlier ME, DeFries RS, Voulgarakis A, et al., (2012) El Niño and health risks from landscape fire emissions in southeast Asia. *Nat Clim Chang* 3:131–136. doi: 10.1038/nclimate1658.

McBride AC, Dale VH, Baskaran LM, et al., (2011) Indicators to support environmental sustainability of bioenergy systems. *Ecol Indic* 11:1277–1289. doi: 10.1016/j.ecolind. 2011.01.010.

McGuire KL, D'Angelo H, Brearley FQ, et al., (2015) Responses of soil fungi to logging and oil palm agriculture in Southeast Asian tropical forests. *Microb Ecol* 69:733–47. doi: 10.1007/s00248-014-0468-4.

MEA (2005) Ecosystems and Human Well-being: Current States and Trends. Island Press.

Meyfroidt P, Lambin EF, Erb K-H, Hertel TW (2013) Globalization of land use: distant drivers of land change and geographic displacement of land use. *Curr Opin Environ Sustain* 5:438–444. doi: 10.1016/j.cosust. 2013.04.003.

Miettinen J, Hooijer A, Shi C, et al., (2012) Extent of industrial plantations on Southeast Asian peatlands in 2010 with analysis of historical expansion and future projections. *Glob Chang Biol Bioenergy* 4:908–918. doi: 10.1111/j.1757-1707.2012.01172.x.

Nepstad DC, Stickler CM, Almeida OT (2006) Globalization of the Amazon Soy and Beef Industries: Opportunities for Conservation. *Conserv Biol* 20:1595–1603.

Nurulita Y, Adetutu EM, Kadali KK, et al., (2014) The assessment of the impact of oil palm and rubber plantations on the biotic and abiotic properties of tropical peat swamp soil in Indonesia. *Int J Agric Sustain* 13:150–166. doi: 10.1080/14735903.2014.986321.

Page SE, Morrison R, Malins C, et al., (2011a) Review of Peat Surface Greenhouse Gas Emissions from Oil Palm Plantations in Southeast Asia. The International Council on Clean Transportation, Washington, D.C.

Page SE, Rieley JO, Banks CJ (2011b) Global and regional importance of the tropical peatland carbon pool. *Glob Chang Biol* 17:798–818. doi: 10.1111/j.1365-2486.2010.02279.x.

Page SEF, Siegert JO, Rieley, et al., (2002) The amount of carbon released from peat and forest fires in Indonesia during 1997. *Nature* 420:61–65.

Phalan B, Bertzky M, Butchart SHM, et al., (2013) Crop Expansion and Conservation Priorities in Tropical Countries. *PLoS One* 8:e51759. doi: 10.1371/journal.pone.0051759.

Phalan B, Fitzherbert EB, Rafflegeau S, et al., (2009) Conservation in oil-palm landscapes. *Conserv Biol* 23:244–5; author reply 245–6. doi: 10.1111/j.1523-1739.2008.01151.x.

Phalan B, Green R, Balmford A (2014) Closing yield gaps: perils and possibilities for biodiversity conservation. *Philos Trans R Soc Lond B Biol Sci* 369:20120285. doi: 10.1098/rstb.2012.0285.

Phelps J, Carrasco LR, Webb EL, et al., (2013) Agricultural intensification escalates future conservation costs. *Proc Natl Acad Sci. doi:* 10.1073/pnas.1220070110.

Pinzi S, Garcia IL, Lopez-Gimenez FJ, et al., (2009) The Ideal Vegetable Oil-based Biodiesel Composition: A Review of Social, Economical and Technical Implications. *Energy & Fuels* 23:2325–2341. doi: 10.1021/ ef801098a.

Prescott GW, Edwards DP, Foster WA (2015) Retaining biodiversity in intensive farmland: epiphyte removal in oil palm plantations does not affect yield. *Ecol Evol* 5:1944–54. doi: 10.1002/ece3.1462.

Proctor S, McClean CJ, Hill JK (2011) Protected areas of Borneo fail to protect forest landscapes with high habitat connectivity. *Biodivers Conserv* 20:2693–2704. doi: 10.1007/s10531-011-0099-8.

Quah E (2002) Transboundary Pollution in Southeast Asia: The Indonesian Fires. *World Dev* 30:429–441. doi: 10.1016/S0305-750X(01)00122-X.

Reddington CL, Yoshioka M, Balasubramanian R, et al., (2014) Contribution of vegetation and peat fires to particulate air pollution in Southeast Asia. *Environ Res Lett* 9:094006. doi: 10.1088/1748-9326/9/9/094006.

Rist L, Feintrenie L, Levang P (2010) The livelihood impacts of oil palm: smallholders in Indonesia. *Biodivers Conserv* 19:1009–1024. doi: 10.1007/s10531-010-9815-z.

Rival A, Levang P (2014) Palms of controversies: Oil palm and development challenges. Center for International Forestry Research (CIFOR).

Sargeant HJ (2011) Vegetation Fires in Sumatra Indonesia. Oil Palm Agriculture in the Wetlands of Sumatra: Destruction or Development.

Sassen M, Sheil D, Giller KE, ter Braak CJF (2013) Complex contexts and dynamic drivers: Understanding four decades of forest loss and recovery in an East African protected area. *Biol Conserv* 159:257–268. doi: http://dx.doi.org/10.1016/j.biocon.2012.12.003.

Savilaakso S, Garcia C, Garcia-Ulloa J, et al., (2014) Systematic review of effects on biodiversity from oil palm production. *Environ Evid* 3:4. doi: 10.1186/2047-2382-3-4.

Sayer J, Ghazoul J, Nelson P, Klintuni Boedhihartono A (2012) Oil palm expansion transforms tropical landscapes and livelihoods. *Glob Food Sec* 1:114–119. doi: http://dx.doi.org/10.1016/j.gfs.2012.10.003.

Sheil D, Casson A, Meijaard E, et al., (2009) The impacts and opportunities of oil palm in Southeast Asia: What do we know and what do we need to know? CIFOR, Bogor, Indonesia.

Singh R, Low E-TL, Ooi LC-L, et al., (2013) The oil palm SHELL gene controls oil yield and encodes a homologue of SEEDSTICK. *Nature* 500:340–4. doi: 10.1038/nature12356.

Smit HH, Meijaard E, van der Laan C, et al., (2013) Breaking the Link between Environmental Degradation and Oil Palm Expansion: A Method for Enabling Sustainable Oil Palm Expansion. *PLoS One* 8:e68610. doi: 10.1371/journal.pone.0068610.

Snaddon JL, Willis KJ, Macdonald DW (2013) Oil-palm replanting raises ecology issues. *Nature* 502:170–171.

Taylor PG, Bilinski TM, Fancher HRF, et al., (2014) Palm oil wastewater methane emissions and bioenergy potential. *Nat Clim Chang* 4:151–152. doi: 10.1038/nclimate2154.

Teuscher M, Vorlaufer M, Wollni M, et al., (2015) Trade-offs between bird diversity and abundance, yields and revenue in smallholder oil palm plantations in Sumatra, Indonesia. *Biol Conserv* 186:306–318. doi: 10.1016/j.biocon.2015.03.022.

Tscharntke T, Tylianakis JM, Rand TA, et al., (2012) Landscape moderation of biodiversity patterns and processes - eight hypotheses. *Biol Rev Camb Philos Soc* 87:661–85. doi: 10.1111/j.1469-185X.2011.00216.x.

Uryu Y, Mott C, Foead N, et al., (2008) Deforestation, degradation, biodiversity loss and CO2 emission in Riau, Sumatra, Indonesia. World Wildlife Fund (WWF), Washington D.C.

Van Dam J, Junginger M, Faaij APC (2010) From the global efforts on certification of bioenergy towards an integrated approach based on sustainable land use planning. *Renew Sustain Energy Rev* 14:2445–2472.

Van der Werf GR, Dempewolf J, Trigg SN, et al., (2008) Climate regulation of fire emissions and deforestation in equatorial Asia. *Proc Natl Acad Sci U S A* 105:20350–5. doi: 10.1073/pnas.0803375105.

Villoria NB, Golub A, Byerlee D, Stevenson J (2013) Will Yield Improvements on the Forest Frontier Reduce Greenhouse Gas Emissions? A Global Analysis of Oil Palm. *Am J Agric Econ* 95:1301–1308. doi: 10.1093/ajae/aat034.

Wich SA, Garcia-Ulloa J, Kühl HS, et al., (2014) Will oil palm's homecoming spell doom for Africa's great apes? *Curr Biol* 24:1659–63. doi: 10.1016/j.cub.2014.05.077.

Wich SA, Gaveau D, Abram N, et al., (2012) Understanding the Impacts of Land-Use Policies on a Threatened Species: Is There a Future for the Bornean Orang-utan? *PLoS One* 7:e49142. doi: 10.1371/journal. pone.0049142.

Wicke B, Sikkema R, Dornburg V, Faaij A (2011) Exploring land use changes and the role of palm oil production in Indonesia and Malaysia. *Land use policy* 28:193–206. doi: http://dx.doi.org/10.1016/j.landusepol.2010. 06.001.

Wicke B, Verweij P, Meijl H van, et al., (2012) Indirect land use change: review of existing models and strategies for mitigation. *Biofuels* 3:87–100.

World Bank (2011) The World Bank Group Framework and IFC Strategy for Engagement in the Palm Oil Sector. International Finance Corporation, Washington DC, USA.

Wösten JHM, Clymans E, Page SE, et al., (2008) Peat–water interrelationships in a tropical peatland ecosystem in Southeast Asia. CATENA 73:212–224. doi: 10.1016/j.catena.2007.07.010.

Wösten JHM, Van Den Berg J, Van Eijk P, et al., (2006) Interrelationships between Hydrology and Ecology in Fire Degraded Tropical Peat Swamp Forests. *Int J Water Resour Dev* 22:157–174. doi: 10.1080/079006 20500405973.

Zen Z, McCarthy JF, Gillespie P (2008) Linking pro-poor policy and oil palm cultivation.

Ziegler AD, Phelps J, Yuen JQI, et al., (2012) Carbon outcomes of major land-cover transitions in SE Asia: great uncertainties and REDD+ policy implications. *Glob Chang Biol* 18:3087–3099. doi: 10.1111/j.1365-2486.2012.02747.x.

In: Monoculture Farming
Editors: Tapan Kumar Nath and Patrick O'Reilly
ISBN: 978-1-63485-166-4

Chapter 5

SWIDDENERS' PERCEPTION ON MONOCULTURE OIL PALM IN EAST KALIMANTHAN, INDONESIA

Daisuke Terauchi** and Makoto Inoue
The University of Tokyo, Tokyo, Japan

ABSTRACT

The objectives of this chapter are 1) describing swiddeners' perceptions on land-use decision-making and land-use strategies in Besiq Village, West Kutai District, where development of large oil palm plantations are planned, and 2) determining the state of oil palm cultivation in Paser District, where oil palm plantations were developed more than 20 years ago, then examining the problems of oil palm cultivation in the eyes of swiddeners.

The people in Besiq Village evaluated land uses such as rattan, rubber, and oil palm cultivation based on three criteria: "return on labor," "flexibility of land and resource use," and "autonomy." They selected rubber cultivation as a source of regular cash income because of the high "return on labor," and also maintained rattan cultivation as a source of cash income for emergencies, and as material for baskets because of its high "flexibility of land and resource use." Villagers hesitated to introduce oil palm plantation development because of low "flexibility of land and resource use" and "autonomy," although they had expectations for a high "return on labor." Therefore villagers agreed to plantation development in the customary forest of the upper river area, but refused it in the vicinity of the village, where land is de facto owned by villagers. This suggests that the villagers adopted a "mosaic livelihood strategy," whereby they try to combine rattan cultivation and rubber cultivation in a complementary fashion, and accept the development of oil palm plantations partly in an effort to avoid social concerns.

Based on the realities of oil palm cultivation in Paser District, the problems of oil palm plantations were examined in terms of the three criteria used by swiddeners. The return on labor of oil palm cultivation was high, but if land was not leveled and roads were not sufficient as in West Kutai District, the return on labor is considered to be almost the same as that of rattan cultivation, and less than that of rubber cultivation.

* Corresponding author address: Minamidai 4-44-4, Nakano-ku, Tokyo, Japan, 164-0014. Email: aa087095 @gmail.com.

Regarding flexibility of land and resource use, swiddeners practiced agroforestry based on oil palms with rubber, teak, fruit trees or medicinal plants, mushroom collection in oil palm plantations, and silvopasture. Our study found that these practices could improve the flexibility of land and resource use. It is also effective for villagers to determine land-use planning through a participatory approach which decides which areas will be conserved and which developed.

There appeared to be no serious problem regarding local autonomy. Companies started paying heed to local people after democratization in 1998. Market monopolization by companies also lessened because there are now more company mills and broker companies, and because there are many more independent people collecting fresh fruit bunches.

It is necessary for the government and companies to recognize the diverse local peoples' desires for land use decision-making and to incorporate them into development plans. It is also necessary to provide local people with useful information about oil palm cultivation to eliminate the difference between their expectations and reality.

Keywords: Swiddeners, perception, oil palm plantation, East Kalimantan

INTRODUCTION

The oil palm cultivation area in Indonesia expanded rapidly from 673,033 ha in 1990 to 6,090,000 ha in 2011 (FAO, 2013). The Indonesian government has promoted oil palm plantation development, and oil palm companies have led plantation development. The development frontier is moving from Sumatra to Kalimantan and Papua (Sheil et al., 2009). Oil palm plantation development has not only caused deforestation but has also brought about profound socioeconomic changes in local societies in the frontier areas. In Kalimantan, which is one of the most typical frontiers, the Dayak people, who are indigenous swiddeners, have long made a living based on forest resources.[10] One would expect that plantation development would drastically change their livelihoods. The main questions this chapter hopes to answer are how swiddeners cope with oil palm plantation development, why they take certain actions to cope, and how this changes their lives. We examine these questions using a case study on West Kutai District, East Kalimantan Province, which has rich tropical rainforests and swiddeners whose lives are based on them, and where large oil palm plantation development is planned.[11]

Most Kalimantan swiddeners have made livings based on swidden agriculture, otherwise known as shifting cultivation, for food production, and also tree-crop cultivation such as rattan and rubber in fallow swidden fields for cash income. These traditional swidden and tree-crop cultivation systems are recognized as well-integrated land use systems (Pelzer, 1978). For example, it is said that the swidden and rattan cultivation system (Inoue, 1991; Weinstock, 1983; Pambudhi et al., 2004; Godoy, 1990) and the swidden and rubber cultivation system (Dove, 1993; de Jong, 2001; Penot, 2004; Pelzer, 1978) are to some extent

[10] "Dayak" is defined as the non-Muslim, non-Malay natives of Borneo Island, including the Indonesian portion called Kalimantan, the Malaysian portion called Sabah and Sarawak, and Brunei. Generally the Dayak live in the interior of Borneo Island (King, 1993).

[11] Since mid-2013, East Kalimantan has been divided into East Kalimantan and North Kalimantan, and West Kutai has been divided into West Kutai and Mahakam Ulu. This chapter uses the pre-2013 administrative boundaries of East Kalimantan Province and West Kutai District.

able to maintain forest ecosystems of rattan or rubber gardens in the same way as fallow swidden fields, to maintain the long-term swidden cycle, and to secure cash income from rattan or rubber gardens in the swidden fallow period. Additionally, these land-use combinations make subsistence activity and market-oriented activity compatible. Swiddeners can therefore cope with market fluctuations and natural disasters such as long droughts (Cramb, 1993; Dove, 1993; Gönner and Seeland, 2002; Gönner 2011).

By contrast, oil palm cultivation is basically a large plantation system. The oil palm fruit bunches, called fresh fruit bunches (FFBs), must be processed in mills within 48 hours to maintain crude palm oil quality (Sheil et al., 2009; Feintrenie et al., 2010a). Local people must therefore collaborate with companies that are capitalized well enough to build the necessary mills, roads, and other infrastructure when oil palm plantations are first being established. Accordingly, oil palm companies need large plantation areas of more than 4000 ha to recover their investments (Sheil et al., 2009). Oil palm plantations have been developed under the PIR (*Perusahaan Inti Rakyat*) scheme by the Indonesian government. Companies establish mills and their own nucleus estates (locally called Inti), as well as smallholdings for participating local people (locally called Plasma). In general, land shares of Inti and Plasma in all estate areas are decided by negotiations between companies and villages, but Inti shares are limited to less than 80% of an estate's total land area, and Plasma shares must be more than 20%.[12] Plasma holdings are distributed to participating villagers, with each receiving 2 ha of Plasma land. Participating villagers organize farmers' groups and cooperatives. They must sell their FFBs to the contracted company's mill for cash income and to pay back loans. Plasma villagers are forced to depend heavily on companies owing to their monopolization of the FFB market. This is behind the large differences between traditional land use and the economic system of swiddeners on the one hand, and those of oil palm plantations on the other.

It is said that large-scale development of oil palm plantations by companies have caused environmental problems such as deforestation, biodiversity loss, and increasing floods, as well as human rights abuses among local populations such as seizing customary lands from local people, heavy crackdowns on local protestors (beatings and even killings) by the companies' security forces, by the police, or by the military, pollution of rivers used by local people for drinking and washing water, and broken promises that companies will employ local people and build facilities such as roads, schools, and irrigation infrastructure for local communities. Other problems are the economic vulnerability of monoculturing, and the loss of local cultures, such as the destruction of sacred forests and the loss of traditional customs and values (Marti, 2008; Sheil et al., 2009). On the other hand, several researchers found that local people obtained higher incomes by oil palm cultivation (Feintrenie et al., 2010a; Kawai, 2011; Kawai and Inoue, 2010; Rist et al., 2010; Zen et al., 2006). It was even reported that people living near companies' mills started to establish their own oil palm plantations and selling their FFBs to raise their incomes (Cramb and Sujang, 2013).

During the 1968-98 rule of President Suharto, oil palm plantations were developed by companies which obtained legal licenses from the government without the agreement of indigenous peoples, but after the Suharto era democratization progressed and oil palm plantation companies have since needed to obtain the agreement of indigenous peoples before they develop a plantation. Because indigenous peoples are now able to decide whether or not

[12] Mentioned in article 11 of the Decree of the Ministry of Agriculture No. 26 in 2007.

to accept plantation development, the future livelihoods of swiddeners depend on the swiddeners' own decisions. It is clear that understanding swiddeners' perceptions of land uses and land-use decision-making, in other words land-use strategies, is important for examining Kalimantan's swidden society in the future. Although the above-cited previous research has shed light on the ecological and economic functions of swidden and tree-crop cultivation systems, and on the impact of oil palm cultivation and production on local environments and on the economic, social, and cultural aspect of local societies, only a few studies (Faintrenie et al., 2010a; Faintrenie et al., 2010b) elucidate local peoples' perceptions of land use and land-use strategies. This chapter therefore describes swiddeners' perception of land use and land-use strategies in West Kutai District of East Kalimantan Province, where large oil palm plantations are in development. The importance of the participatory method has become increasingly recognized in the current practical and academic fields of rural development (Chambers, 1983; 1997). Given that it is essential to incorporate local peoples' hopes and concerns into rural development projects, this study could contribute to and improve future participatory rural development projects.

We identify the important considerations regarding swiddeners' perception on land use and land-use strategy in West Kutai District, and then ascertain whether oil palm cultivation fulfills these considerations by examining a case from Paser District of East Kalimantan, where oil palm plantations were developed more than 20 years ago.

METHOD

Selection and Description of Research Sites

East Kalimantan Province

The province of East Kalimantan is the center of forest cover in Indonesia, accounting for 11% of the national total (DK, 2008). Table 1 shows the state of oil palm plantation development in East Kalimantan in 2012. The total developed area of oil palm plantations occupies 18.1% (961,797 ha) of the province's non-forestry zones, where the government can give agricultural development right to companies.[13] Of the total oil palm plantation area, 76.4% (735,077 ha) is owned by companies and 23.6% (226,720 ha) is owned by local people. The area planned for development as oil palm plantations by companies, in other words an undeveloped block in the area covered by a location permit (*Izin lokasi*), which is the first permit granted to a developer by the government, accounts for 57.5% (3,061,982 ha) of the province's non-forestry zones.[14] This means that there are still many potential areas that could be developed as oil palm plantations by companies. Most of the residential and forest land of the Dayak people is located within non-forestry zones. Consequently, areas intended for oil palm cultivation overlap Dayak residential land.

---

[13] Indonesian land is classified as "forestry zones (*Kawasan Budidaya Kehutanan*)" and "non-forestry zones (*Kawasan Budidaya Non Kehutanan*)" by law. The development of oil palm plantations is permitted within "non-forestry zones." These classifications, however, do not indicate the land cover of a region. For example, in the non-forestry zones, there are landscapes rich in natural forests and agroforestry landscapes that have been developed for traditional land/forest uses by indigenous swiddeners.

[14] Oil palm companies must first get location permits (*Izin Lokasi*) and then plantation business permits (*Izin Usaha Perkebunan*) from the government, in that order, to develop oil palm plantations.

Table 1. Oil palm plantation development in research sites in 2012 (ha)

	East Kalimantan Province	West Kutai District	Paser District
Non-forestry zones	5,324,488	932,266	609,015
Total oil palm plantation areas	961,797	58,818	157,116
Owned by companies	735,077	55,314	85,976
Owned by local people[1]	226,720	3504	71,140
Area under development plan by companies	3,061,982	569,190	235,276

[1] Including Plasma areas and non-Plasma areas that are established by local people without collaboration with companies.

Source: DPPKT (2013) and list of oil palm plantation companies in East Kalimantan in 2014.

West Kutai District and Besiq Village

West Kutai District is located on the upper Mahakam River, which is one of the main rivers in East Kalimantan Province (Figure 1). Because it is one of the most-inland districts, it has more remaining natural forests than other districts. West Kutai District has a population of 184,394 and a land area of 3,162,870 km^2 (BPS, 2013). Population density is 5.83 persons/km^2. Dayak people account for 65.9% of the population (Gönner et al., 2007).

Table 1 also shows the state of oil palm plantation development in West Kutai District in 2012. The total developed area of oil palm plantations covers 6.3% (58,818 ha) of the district's non-forestry zones. Companies own 94.0% (55,314 ha) of oil palm plantation land and local people own 6.0% (3504 ha). Land under development plans for oil palm plantations by companies is 61.1% (569,190 ha) of the district's non-forestry zones. This indicates that oil palm plantation development by companies covers much of the district's land, and that companies still have much room for additional plantation development there. The center of oil palm cultivation and the plantations of companies and of local people are in Tanjung Isui, Muara Pahu, and Jempang subdistricts, which are located in the Mahakam River's downstream region. In other subdistricts, oil palm cultivation by companies and local people has not yet started, and oil palm plantation development has just begun.

Besiq Village, which is the focus of this chapter, is located the farthest up the Kedang Pahu River, a tributary of the Mahakam River. In 2010 the population was 1063 and there were 282 households. The main ethnic group is the Benuaq people, a Dayak subgroup. The land around Besiq Village is old secondary forest, rattan gardens, and traditional fruit gardens. Mature forests still remain upriver of Besiq Village.

This village is a famous rattan cultivation site (Pambudhi et al., 2004). Villagers practice swidden agriculture and cultivate rattan in swidden fallow fields. This has been recognized as a sustainable cultivation system in the sense of maintaining the long-term swidden cycle and generating cash income in swidden fallow periods (Inoue, 1991; Inoue and Lahjie, 1990). However, rattan farm gate prices have stagnated since the 1990s even though commodity prices are on the rise (Table 2). This is worsening the economy in Besiq Village.

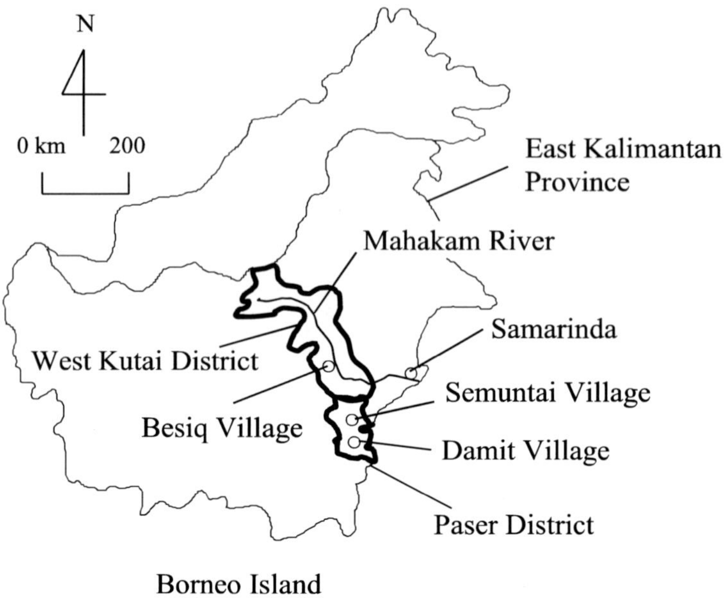

Figure 1. Map of research site (created by authors).

The government implemented two rubber plantation projects, namely PRPTE (*Peremajaan, Rehabilitasi dan Perluasan Tanaman Ekspor*) in the 1980s and TCSSP (*Tree Crop Smallholder Sector Project*) in the 1990s, near the capital city of West Kutai District to improve the local economy. Around 2000, villagers in Besiq Village, who were influenced by these rubber plantation projects, started to plant rubber tree seeds and seedlings in fallow swidden fields.

Two oil palm plantation companies in 2004 and three oil palm companies in 2008 started procedures to obtain rights for plantation development from the government. All areas of the non-forestry zone around Besiq Village came within the planned area of oil palm plantation development. Besiq Village is suitable as a research site to examine the local peoples' perceptions of land use and land-use strategies. Further, people in Besiq Village badly need information on oil palm cultivation.

Table 2. Changes in commodity prices in Besiq

	Nov. 1988	Jan. 2010
Rice (IDR/kg)	660	7400
Sugar (IDR/kg)	1000	12,000
Salt (IDR/400g)	150	2000
Gasoline (IDR/L)	700	7000
Rattan (IDR/kg)	250-750	1000
Exchange rate (US$1)	IDR 1731	IDR 9292

Sources: Field research in 2010 and Inoue (1989).

Damit Village and Semuntai Village in Paser District

Paser District is located in the southern part of East Kalimantan Province (Figure 1). Its population is 247,612 and its land area is 11,487 km^2 (BPS, 2013). Population density is 21.56 persons/km^2. The Pasir are the district's indigenous people[15].

Oil palm plantation development in Paser district started in 1983. Table 1 shows the state of development there in 2012. Total developed oil palm plantation area occupies 25.8% (157,116 ha) of the district's non-forestry zones. Companies own 54.7% (85,976 ha) of the oil palm plantation area and local people own 45.3% (71,140 ha). Many local people have self-established oil palm plantations as well as Plasma fields from participation in the PIR scheme. Land subject to oil palm plantation development plans of companies accounts for 38.6% (235,276 ha) of the district's non-forestry zones. Such land in Paser District is much less than that in West Kutai District.

We selected two research sites: Damit Village and Semuntai Village. Their villagers have a long history of and much experience in oil palm cultivation with the national oil palm company in Paser District.

In 2009, the population of Damit Village was 3161 and there were 811 households. Oil palm plantations have been developed here by the national oil palm company since 1991. There were 1500 ha of the national oil palm company's Inti and 752 ha of villagers' oil palm plantations in 2009 (Kawai and Inoue, 2010). The main ethnic group is the Pasir. Most villagers have oil palm plantations, and their main means of livelihood is oil palm cultivation. In addition, some villagers practice paddy cultivation and fishing for self-consumption because Damit Village is on the Kandilo River.

Sumuntai Village is the other research village in Paser District. In 2015, the population of Semuntai Village was 4292 and there were 1253 households. Oil palm plantations have been developed here by the national oil palm company since 1981. The total Inti area of the national and private companies is 3400 ha, and villagers' oil palm plantations total 1275 ha. Ethnically, Semuntai Village is half Pasir and half non-Pasir people who came from elsewhere. The main livelihood means of villagers is oil palm cultivation and paid labor for the national oil palm company, such as harvesting or management of Inti and mill workers.

Data Collection

Field research was conducted between August 2006 and March 2015 in East Kalimantan. The total research period was about two and a half years.

The data used to determine swiddeners' perceptions of land use and land-use strategies in Besiq Village were collected mainly in 10 months between August 2006 and 2008.

There are four neighborhood groups (*Rukun Tetangga*) in Besiq Village. We selected the first group as representative of Besiq Village. The group has 43 households. We randomly selected 15 households from among them and conducted semi-structured interviews regarding villagers' perceptions of rattan cultivation, rubber cultivation, and oil palm cultivation in 2006 and 2007. Following this data collection, we randomly selected 20 households from among

---

[15] The Pasir people originally belonged to the Dayak but converted to Islam (King, 1993). Although the Pasir are separate from the Dayak in this sense, they have many similarities with the Dayak. Their language is similar to the Benuaq language and they also practice swidden agriculture and hunting.

the 43 households and conducted semi-structured interviews regarding people's hopes and concerns about ongoing plans for oil palm plantation development by companies in 2008.[16]

The data on the state of oil palm cultivation in Paser District were collected during one week in 2009 and two weeks in 2015. We conducted interviews with a village chief, several leaders of neighborhood groups, officers of the national oil palm company, leaders of Plasma cooperatives, and officers of the forestry, plantation, and livestock services of Paser District.

SWIDDENERS' PERCEPTIONS ON LAND USE AND LAND-USE STRATEGIES[17]

In this section, we first describe the swiddeners' land-use system in Besiq Village, including rattan and rubber cultivation as well as swidden agriculture, and then describe the swiddeners' perceptions of rattan and rubber cultivation. Second, we describe the oil palm cultivation system and swiddeners' perceptions of oil palm cultivation. Finally we summarize the swiddeners' perceptions of land uses and land-use strategies.

Swidden and Tree-Crop Cultivation System

Swidden Agriculture

Table 3 shows the sequence of operations in swiddens cultivated by villagers in Besiq Village. Swidden agriculture involves cutting down trees, drying, burning, dibbling, and sowing rice, then weeding and harvesting the rice. It is important to note that the sequence of operations from cutting to burning must be finished by the rainy season, or a good rice harvest cannot be expected. When villagers start cutting and felling trees depends on forest conditions. Operations are started early if there are many big trees. Harvesting runs from late February through April. Usually villagers produce rice and some vegetables for self-consumption.

Benuaq society has several types of work organization. *Kerja Edotn*, *Pelo*, and *Upaah* are often adopted in tree cutting and sowing operations. *Konokng* and *Pelo* are often adopted in harvesting operations, and *Awaat/Perawaat* is often adopted in burning operations. *Awaat/Perawaat* and *Upaah* are also used for everyday tasks. *Betusa* is adopted in rattan and rubber cultivation.

Villagers create swiddens in semi-primary forests, called *Bengkar* in the Benuaq language, that have never been used as swiddens but have traditionally been used for harvesting small amounts of timber or non-timber forest products. Swiddens can also be created in secondary forests or tree crop gardens, called *Krat*, that were established by villagers or their ancestors. Villagers usually change their swidden sites every year. Generally they plant seeds or seedlings of tree crops, such as rubber, rattan, and fruit, at the same time as planting rice, or immediately after harvesting rice.

---

[16] Due to the semi-structured interview, Table 12 includes the responses of only 14 of the 20 informants.

[17] This section is based on a modification of Terauchi (2010), Terauchi et al. (2010), and Terauchi (2013).

Table 3. Annual swidden task schedule

Month	local Name	Operation	Worker gender	Work organization[1]
May - Sep	Nokap	Cutting small trees	Men and Women	a, b, g
	Noang	Felling big trees	Men	a, b, g
	Nutu Joa	Removing branches of fallen trees	Men	a
	Oikng Joa	Drying trees in the sun	-	-
	Nyuru	Burning trees	Men	a, c, d
	Mongkakng	Re-burning unburned trees	Men	a
Oct	Ngasak	Dibbling and sowing rice	Men and Women	a, b, g
Oct - Jan	Ngejikut	Weeding	Women	a
Feb - Mar	Ngotapm	Harvesting ripe rice	Men and Women	a, b, e

Note 1:

a. Kerja Edotn: Individual family labor system.
b. Pelo: Labor exchange among households characterized by direct parity reciprocity.
c. Awaat: Spontaneous offer of labor without compensation.
d. Perawaat: Offer of labor based on request without compensation.
e. Konokng: A kind of paid labor system in harvesting rice, where remuneration is 1 Amoq basket (20-30 kg) of rice per day or the amount of rice harvested from noon to evening.
f. Betusa'(Menguli): A kind of paid labor system in harvesting rattan or rubber, where remuneration is usually 50% of the total harvest.
g. Upaah: A paid labor system.

Source: Modified from Terauchi (2010) and Terauchi and Inoue (2011).

Rattan Cultivation

Although there are numerous species of rattan, we chose to focus on the Sega variety (*Calamus caesius*), which is the foremost species of rattan cultivated not only in Besiq Village, but also in East Kalimantan. Sega rattan is cultivated using the following steps.

Generally, villagers collect rattan seeds in their own rattan gardens or from natural forests. Villagers plant rattan seeds rather than seedlings after harvesting rice or at the same time as planting rice because rattan seeds can be gathered easily. Rattan seeds may be broadcast, but usually the planting interval is around 4 m by 4 m. Villagers then weed around the rattan seedlings (*Nokap*). After the forest canopy has closed, they begin girdling the big trees (*Ngeleleng*) around the rattan to kill them so that the rattan gets nutrients and sunlight.

Nokap and *Ngeleleng* are carried out irregularly and extensively. Useful trees, such as fruit and wood species, are retained and occasionally planted in the garden. Customary trees in which honeybees nest must be retained so that villagers can harvest honey from them. Rattan garden landscapes resemble forests (Photo 1).

The first harvest (*Pikat we*) is about 8 -10 years after planting. Subsequently, rattan canes that regrow from the same stump can be harvested every 3–4 years. The harvest is difficult and painful due to the many thorns. Villagers cut the rattan loose up in the trees, and then cut it at a height of 1–1.5 m above ground to regrow new rattan canes from the same stump.

Photo 1. Landscape of a rattan garden (Source: Terauchi (2010) and Terauchi (2013)).

Table 4. Utensils made with rattan

Local name	Utensil	Usage and characteristics of goods
Gawakng Lengkakng	Basket	Basket used for carrying everyday items; Coarse reticulation and soft basket
Gawakng Anyet	Basket	Basket used in daily life; fine reticulation; design made by weaving with black rattan
Barangka	Basket	Basket for carrying large, heavy things such as firewood and cassava
Kiang	Basket	Basket attached to a board to protect back from durian thorns, blood of wild animals, etc.
Bisakng	Basket	Basket for bringing rice seeds to sow in swidden field; attached to one's waist
Gamaq	Basket	Basket for keeping harvested rice; attached to one's waist
Amoq	Basket	Basket for carrying harvested rice; harvested rice put in Gamaq is transferred to Amoq
Apai jaliq	Mat	Parallel weave mat
Apai we	Mat	Diagonal weave mat
Putatn	Birdcage	Cage for keeping pet birds
Buuq	Fishing basket	Basket used as fish trap. Made of bamboo and rattan

Source: Modified from Terauchi et al., (2010) and Terauchi (2013).

Finally the cut lengths of rattan are pulled down to the ground while removing the thorns. Villagers cut the harvested rattan into 6-m lengths and make bundles weighing 30–40 kg, which are then carried from the forest to the house of a rattan collector, who pays the villagers in cash. If harvested rattan is left in the forest for a long time, weathering changes its color and lowers the price. Carrying rattan across rugged terrain and several small rivers is arduous and dangerous. One person can usually harvest around 100 kg/day. Although there are more than 100 kg of rattan per ha, one day is not enough time for a person to harvest more than 100 kg. In one year a productive rattan garden yields 2500–4000 kg/ha. The price paid to villagers in 2011 was IDR 1300/kg. Sometimes rattan garden owners adopt *Betusa'* when they need labor or workers need jobs. Villagers hired as workers harvest rattan in others' rattan gardens, with income being split fifty-fifty. In addition to sale, rattan can be used by the producers for various practical purposes such as rope and baskets (Table 4). In addition, because villagers manage useful trees in their rattan gardens, they can harvest fruit, firewood, honey, and building timber in addition to rattan (Table 5).

Table 5. Tree species planted or left in rattan and rubber gardens[1]

English name	Scientific name	Usage
Menggeris[2]	*Koompassia excelsa* (BECC.) TAUB.	Harvesting honey[4]
Borneo ironwood	*Eusideroxylon zwageri* TEIJSM. et BIM.	Building material
Meranti[3] Bangkirai Tengkawang	*Shorea* spp.	Building material
Kapur[2]	*Dryobalanops* spp.	Building material
Borneo Teak	*Intsia bijuga* OK.	Building material
Lambutan	*Nephelium lappaceum* L.	Harvesting fruit
Jack-fruit tree	*Artocarpus heterophyllus* LAM.	Harvesting fruit
Kapul[2]	*Baccaurea macrocarpa* (Miq.) M.A.	Harvesting fruit
Cempadak[2]	*Artocarpus integer* (Thunb) Merr.	Harvesting fruit
Asam[2]	*Mangifera pajang* Kastern.	Harvesting fruit
Durian[5]	*Durio* spp.	Harvesting fruit; trunks for coffin material
Areca-nut palm	*Areca catechu* L.	Harvesting fruit; leaves for decorative material in rituals

[1] This table is based on interviews with villagers. Based on interviews with local people and on vegetation surveys, Gouyun et al., (1993) found that 23 species (in South Sumatra) and 26 species (in Jambi) of useful trees, excluding rubber trees, growing in rubber gardens established by local people are managed and used by the local people in those two regions.

[2] Indonesian names used because there are no English names.

[3] Tree name given by local people is used because it was difficult to identify specific tree species.

[4] Under customary rules, menggaris trees, which host honeybee hives, are treated as customary trees and may not be cut. Other tree species are also treated as customary trees if there are honeybee hives in them.

[5] There are various durian tree species. The trunk of the durian species *Durio zebethinus* MURR is used to make coffins.

Source: Based on Terauchi et al., (2010) and Terauchi (2013).

Villagers harvest rattan cane little by little from one rattan garden or in rotation from several rattan gardens to maintain a cash income flow. A rattan garden can be harvested for several decades. It is said that some rattan gardens have lasted for around 100 years (Inoue, 1991). If the productivity of a rattan garden declines or there are particular reasons such the need for food production or the desire to convert a rattan garden into a tree-crop garden, villagers make a swidden of the rattan garden for food production and establish a new tree garden by replanting a fallow swidden.

Rubber Cultivation

Rubber (*Hevea brasiliensis*) gardens are established in a traditional manner by preparing non-certified seeds/seedlings (non-certified meaning those not of improved varieties), planting, and weeding.

Villagers obtain rubber tree seeds and seedlings from relatives' gardens in Besiq Village or neighboring villages. Fertilizers and herbicides are not used. They plant seeds or seedlings after harvesting rice or at the same time as planting rice. The planting pattern is grid-like because villagers learned the planting method from the rubber plantation projects that were implemented in villages around capital city of West Kutai District, but the pattern is not as strict as the project, and denser than that used in rubber plantations (4 m by 5 m, 500 trees/ha). Rubber gardens need more intensive management (weeding and cutting) than rattan gardens to withstand competition from wild plants and other trees. Rubber seedlings are damaged by wild animals such as deer, but there are no such problems with rattan gardens. Villagers cope with this problem by setting many traps called *Sekokoi* around rubber fields, but the traps are not very useful. Useful trees, such as fruit or wood species and customary trees, are retained and occasionally planted in the same garden. Rubber garden landscapes therefore look like forests (Photo 2).

Rubber trees can occasionally be tapped in 5–6 years depending on their growth or diameter, but harvesting generally begins after 7 years. Latex is harvested 4–5 days per week. It is said that trees must not be tapped some days to maintain productivity, and harvesting cannot be conducted on rainy days because sap that has not yet solidified overflows from harvesting cups. It takes an average of 3–4 hours to tap 1 ha of rubber trees starting in the early morning (between 5 and 9 a.m.). The tasks of tapping and carrying latex are not backbreaking, dangerous, and physically demanding like rattan harvesting . Harvested rubber is sold to rubber collectors in Besiq Village or to outside collectors. The average amount of rubber harvested is 24.2 kg/ha/week for the 7th through 25th years, while the production peak is achieved from the 15th to 18th years. The average farm gate price of rubber paid to villagers in 2011 was IDR 11,200/kg. Sometimes rubber garden owners adopt *Betusa,'* and in that case owners and workers split the proceeds fifty-fifty. In addition to latex, villagers can also harvest fruit, firewood, wild animals, and timber for houses (Table 5). Unlike rattan, latex has only limited uses. Branches of rubber trees are used as firewood, rubber seeds as pig feed, and rubber gum as birdlime.

Villagers can harvest latex from the same trees for 30 years or more. They can convert old non-producing rubber gardens to swiddens for food production, and later establish new tree gardens by replanting in swidden fallow fields.

Photo 2. Landscape of a rubber garden (Source: Terauchi (2013)).

Perception on Rattan and Rubber Cultivation

Villagers' Evaluation of Rattan Cultivation

Villagers mentioned that the advantages of rattan cultivation are being able to harvest much at one time, extensive management is sufficient, the many practical uses of rattan, harvests are not limited to a certain time of year, and rattan can be readily sold for cash (Table 6).

Table 6. Advantages of rattan cultivation (n = 15)[1]

Harvest much at one time	13
Extensive management	5
Various daily uses of rattan	5
No limitation on harvesting time	3
Easy to sell rattan for cash income	1

[1] Multiple answers are allowed.

Source: Based on Terauchi et al., (2010) and Terauchi (2013).

Villagers can readily obtain cash income in an emergency such as a family member's illness because much can be harvested at one time, any time of the year. Continuous cash income is possible because rattan can be harvested year-round, harvesting can be done little by little in one rattan garden or in rotation among several gardens, and *Betusa'* working organization can be used. Although villagers did not mention price stability, it is an important factor related to the advantages mentioned above. The price of Sega rattan in Besiq Village was stable even after the Lehman Brothers crash in late 2008.

On the other hand, the disadvantages are a long interval (4 years) between harvests, the heavy labor of carrying rattan, the painful and physically demanding harvest operation, and the deterioration of rattan when stored for a long time (Table 7).

Villagers cite the long interval between rattan harvests as a disadvantage because they compare rattan harvesting with rubber harvesting, which can be conducted almost every day. But villagers cope with this disadvantage with harvesting techniques and by using *Betusa.'* Harvesting and transport difficulties underlie villagers' low preference for rattan cultivation.

Table 7. Disadvantages of rattan cultivation (n = 15)[1]

Long interval to next harvest	13
Rattan is heavy to carry	12
Painful and troublesome harvest operation	11
Once ratten is harvested, it does not keep well	1

[1] Multiple answers are allowed.

Source: Based on Terauchi et al., (2010) and Terauchi (2013).

Villagers' Evaluation of Rubber Cultivation

Villagers said that the advantages of rubber cultivation include being able to harvest every day, easy harvesting (tapping) operation, the fact that rubber keeps well in storage, ease of transport, and high price (Table 8).

Rubber harvesting is easy even for women, the elderly, and children, but men usually perform the physically demanding and painful harvesting of rattan and its transport. Villagers note that rubber brings high prices even though it is easy to harvest and transport. Villagers can sell rubber to collectors whenever they want, such as when prices are high or when villagers need cash income, because rubber keeps well for a long time.

On the other hand, the disadvantages are damage by wild animals and insects when trees are immature, the need for intensive management, limited harvest volume per day, the need for fertilizers, and the difficulty of planting other crops or useful trees (Table 9).

Table 8. Advantages of rubber cultivation (n = 15)[1]

Can harvest everyday	14
Easy harvesting operation	13
Rubber keeps well after harvest	11
Easy to transport	11
High price	1
Emerging paid labor opportunity	1
Seeds can be used as pig feed	1

Note 1: Multiple answers are allowed.

Source: Based on Terauchi et al., (2010) and Terauchi (2013).

Table 9. Disadvantages of rubber cultivation (n = 15)[1]

Animal and insect damage	13
Intensive management (high labor input)	6
Limited harvest volume per day	6
Fertilizer needed	1
Difficulty of planting other crops and trees	1

[1] Multiple answers are allowed.
Source: Based on Terauchi et al., (2010) and Terauchi (2013).

The management of rubber gardens, including prevention of animal and insect damage, intensive management, and fertilizer application, was seen as a disadvantage. Animal damage and insect infestations in particular are serious problems. Some people have lost all their rubber trees to wild deer. Villagers have yet to find effective ways to prevent this. The amount of rubber harvested per day or per ha is limited to about 5.4 kg. Rattan on the other hand is limited not by volume but by labor. Some people observed that it is difficult to interplant crops or useful trees with rubber trees because they fear that interplanted crops or trees will compete with rubber trees for nutrients and sunlight. Other people said fertilizers are needed, but almost all villagers interplant and manage rubber gardens without fertilizers. Rubber price instability has a major impact on villagers' lives, although they did not mention this. After the Lehman Brothers crash of late 2008, rubber collectors in Besiq Village and outside collectors stopped buying rubber from villagers.

Summary of Perceptions on Rattan and Rubber Cultivation

Villagers see pros and cons associated with rattan and rubber cultivation. In their view, rattan cultivation involves easy garden management but difficult harvesting and transport, while rubber cultivation involves difficult garden management but easy harvesting and transport. Villagers are willing to increase their rubber gardens (Table 10), and said that rubber cultivation is their main source of cash income.

Table 10. Villagers' intentions for gardens (n = 15)

	Rattan	Rubber
Increase	1	14
Convert	1	0
Maintain	13	0
Undecided	0	1

Source: Modified from Terauchi et al., (2010) and Terauchi (2013).

Villagers chose rubber over rattan because rubber cultivation has a high "return on labor." The concept of return on labor is different from "labor productivity," which means the yield produced by labor in a certain amount of time. Return on labor includes labor characteristics such as heavy/light, dangerous/safe, and physically demanding/easy, as well as labor time. The labor productivity of rattan cultivation is IDR 130,000/man-day while that of rubber cultivation is about IDR 108,400/man-day.[18] The labor productivity of rattan cultivation is higher than that of rubber cultivation, yet villagers chose rubber cultivation as their main cash income source because they considered the easy harvesting and transport of rubber to be advantages, and the physically demanding harvest and transport of rattan to be disadvantages. Although villagers acknowledged that rubber gardens are difficult to manage, hardly any villagers practice intensive management such as frequently guarding against wild animal damage, or using fertilizers, herbicides, and pesticides.[19] If labor productivity were the most important criterion for villagers, one would expect them to harvest all available rattan in a year, but instead, almost all of the villagers hesitate to harvest rattan, and they leave available rattan in their gardens because of the physically demanding harvest, the arduous and dangerous transport, and the low price of rattan. This suggests that villagers prefer rubber cultivation owing to its return on labor rather than its labor productivity.

On the other hand, villagers maintain existing rattan gardens instead of converting them to other gardens (Table 10). They recognized the advantages of rattan cultivation, such as being able to harvest much at one time, and the various practical uses of rattan. Rubber cultivation does not have these advantages. Villagers saw rattan cultivation not as a regular cash income source, but as a cash income source in emergencies, and as materials for baskets or practical utensils. Some people put off harvesting their rattan gardens until the price of rattan increases.

From this one can conclude that villagers combine rattan cultivation and rubber cultivation because they complement the advantages and disadvantages of each other. This can be called a "hybrid land-use strategy."

Oil Palm Cultivation and Swiddeners' Perceptions

Oil Palm Plantation Development Schemes[20]

Oil palm plantations have been developed under the PIR scheme mentioned above. There are actually several types of PIR schemes. We explain the PIR-KKPA (*Kredit Koperasi Primer untuk Anggota*) and PIR-PSM (*Pola Satu Manajemen*) here because these schemes involve the local people of West Kutai District in oil palm cultivation. The PIR-KKPA scheme is based on contracts among a company, participating villagers in PIR-KKPA cooperatives, and banks. The main features of contracts for villagers participating in PIR-KKPA are as follows: 1) Participating villagers organize farmers' groups and cooperatives; 2) the contracted company establishes the Plasma and manages it intensively until the oil palms are four years old, which is an immature stage; 3) establishment and management costs are included in a loan made to the villagers; 4) after Plasma quality has been evaluated, the company cedes 2 ha of Plasma to each villager; 5) villagers manage the Plasma, harvest FFBs from their own Plasma fields, and sell the FFBs to the contracted company's mill; 6) villagers

[18] The labor productivity of rattan cultivation is calculated by multiplying the rattan price (IDR 1300/kg) by 100 kg/day, and that of rubber cultivation is calculated by multiplying the rubber price (IDR 11,200/kg) by 24.2 kg/week per 2.5 manday (half of four or five work days/week). For rattan and rubber, there are almost no capital inputs such as fertilizer, herbicides, or pesticides in management and harvesting.

[19] Some villagers said, "We plant rubber seeds, but whether the seeds grow into productive rubber trees depends on luck."

[20] The explanation of the PIR scheme is based on Kawai and Inoue (2010) and Kawai (2011). See these works for detailed explanations.

are paid in cash by the company, but 30% of FFB income is deducted as loan payments; and 7) villagers can obtain title to their land after finishing repayment.[21]

PIR-PSM began in 2007 as a national agricultural development programme called the "Estate Revitalization Program (*Program Revitalisasi Perkebunan*)." The difference between PIR-KKPA and PIR-PSM is that, in the latter, contracted companies manage Plasma fields and harvest FFBs for 25 years of the oil palm cultivation cycle as representatives of participating villagers in order to maximize FFB quantity and quality. Contracted companies distribute income from Plasma fields to participating villagers. Villagers may work on companies' plantations as employees. Participating villagers can choose between PIR-KKPA or PIR-PSM, but private and national companies generally prefer that villagers choose PIR-PSM, and encourage them to do so. There are no PIR-PSM Plasma gardens which could be harvested in East Kalimantan until 2014.

Oil Palm Cultivation in Plasma Fields

Seedlings are raised for one year and land is cleared before planting seedlings. Around 130 seedlings are planted per ha. After planting, companies perform the labor- and capital-intensive management of applying herbicides, pesticides, and fertilizer, resulting in plantation landscapes (Photo 3). Except for customary trees, other useful trees such as fruit and wood species are not left in the plantations. Even oil palm plantation companies are prohibited from cutting customary trees by customary village rules.

Photo 3. Landscape of oil plam plantation (Source: Terauchi (2010)).

[21] In many cases villagers have only de facto ownership of their land. If villagers participate in PIR oil palm plantation development and finish repaying their Plasma loans, they can obtain title to their Plasma fields.

Villagers perform management and harvesting in their own Plasma fields under the PIR-KKPA scheme after oil palms are four years old. Contracted companies instruct them to apply fertilizers and herbicides available at cooperatives to which the villagers belong, or at other, non-cooperative shops. If villagers buy these agricultural supplies from the cooperatives, the purchase costs are deducted from cash income which contracted companies pay to villagers through cooperatives and farmers' groups. FFBs can be harvested once every two weeks, and the job is usually performed by men. They cut FFBs, which grow at the tops of trees, using long poles that are equipped with hooks. A FFB weighs 20–40 kg. Villagers who are members of the same farmers' groups, which are organized under cooperatives, harvest FFBs on the same day. In a well-managed plantation the estimated average FFB yield is 23.7 t/ha in the mineral soil of Kalimantan (Pahan, 2010). But according to data of the East Kalimantan plantation service, the average FFB yield in East Kalimantan in 2012 was about 15.3 t/ha (DPPKT, 2013).[22] Although FFB prices change depending on the age of oil palm trees, the average price paid to villagers in 2011 was IDR 1353/kg. Villagers who are members of the same farmers' groups harvest FFBs and ship them together in trucks to the contracted company's mill. Contracted companies pay cash through cooperatives and farmers' groups, which then distribute cash income to each participating villager in proportion to the amount of harvested FFBs, and deduct the costs of purchased agricultural supplies.

Usually oil palm cultivation can continue until oil palms are 25 years old. The production peak is eight to 13 years (Pahan, 2010). FFBs are produced for sale alone, not for self-consumption. Other forest products cannot be harvested from oil palm plantations. Villagers do not know whether they can practice swidden on an oil palm plantation after the plantation's 25th year.

Perceptions of Oil Palm Cultivation

Although villagers have no cultivation experience, they are familiar with the oil palm cultivation system from explanatory meetings held by companies and the government, and they get information from villagers who live in Tanjung Isui, Muara Pahu, and Jempang subdistricts, which are downstream from West Kutai District.

Villagers are concerned about drastic lifestyle changes caused by large-scale oil palm plantation development. Such ongoing plantation development induces villagers to adopt different land-use systems and household economies. Villagers have supported themselves with swiddens for food production and rattan and rubber cultivation for cash income. They can harvest various non-timber forest products such as fruit, firewood, and medicinal plants for self-consumption, as well as rattan and rubber from rattan and rubber gardens. They can also use these gardens for swidden fields again and plant new tree crop gardens, depending on the market prices of tree crops. The swidden and rattan/rubber cultivation system reconciles the subsistence economy with the market economy. However, large-scale oil palm plantation development does not allow diverse and flexible land and resource use in the way that the existing land-use system does. Forests for swidden agriculture and existing rattan and rubber gardens are converted to monoculture plantations. Villagers see that it is difficult to use oil palm plantations for swidden fields, and at the same time to harvest forest products to meet daily needs such as firewood while also harvesting oil palm fruit bunches. There is no doubt

---

[22] In 2012 the average FFB yield in villagers' oil palm plantations was 14.6 t/ha, that in private companies' plantations was 15.4 t/ha, and that in national companies' plantations was 17.2 t/ha (DPPKT, 2013).

that the current household economy will be changed to a market economy. This situation also involves not only technical problems with the use of land and resources in large-scale oil palm plantation development, but also social constraints. Companies instruct villagers how to manage plantation production, and the contracts with companies do not permit villagers to interplant or to convert oil palm plantations to other garden types.

Villagers also fear that their livelihoods will be jeopardized if large-scale monoculture oil palm cultivation, upon which the local economy is highly dependent, collapses due to low FFB prices, disease, insect damage, or other causes. These problems are not critical in the existing diverse and flexible land and resource usage system because there are alternatives.

In addition, they worry about their livelihoods when oil palms are still immature (up to four years old) because they have no land or resources to make a living during that period. One oil palm company told villagers that they can work and obtain cash income as plantation workers or mill workers, but villagers doubt that companies will actually employ them as workers, and they worry that companies will require hard labor or even forced labor.

Villagers have concerns about collaboration with oil palm companies because they are highly suspicious of those companies owing to their experience in the Suharto era with the operations of logging companies in the river basin above Besiq Village, which did not benefit villagers' livelihoods but instead deprived them of large trees in natural forests. Villagers came to recognize that companies do not intend to benefit villagers' livelihoods. Similarly, villagers fear that oil palm companies will force villagers to work under severe conditions in Plasma fields as well as Inti. Villagers also fear that companies will refuse to purchase their FFBs or buy them at an unfairly low price because contracts with companies limit villagers in how they can ship FFBs. Companies have villagers over a barrel because villagers make a living from oil palm cultivation alone, and FFBs must be processed within 48 hours at contracted companies' mills.

Some villagers hope to cultivate oil palms themselves without collaborating with companies, but they think this is difficult because they lack the knowledge and techniques of oil palm cultivation, and agricultural supplies such as seedlings, fertilizer, and herbicides.

At the same time, villagers have expectations for oil palm cultivation, such as high cash income, low harvesting frequency (twice a month), promising consumer demand, and stable income from employment at companies' plantations and palm oil mills. Some villagers hope that a stable and high cash income from oil palm cultivation or employment at those companies will give them new lifestyles in which they can purchase all their needs instead of growing food and other needs as they do now.

In view of the foregoing we conclude that villagers see plans for large oil palm plantation development as a high-risk, high-return investment. They have both expectations and concerns for it, but were unsure of what to do. Ultimately, villagers decided to accept the development of oil palm plantations by a company in their customary forest of the upper river area, called *Bengkar* in the local language, where villagers have never practiced swidden, and do not yet own the land (Table 11). The reasons for this decision were that villagers expected high profits from the plantations, and that it was difficult for villagers to use that area as swidden fields or tree gardens due to difficult access. However, they decided to reject plantation development on land which they can use, such as near the village or near rivers and roads, because they want to avoid the problems noted above. Most of these lands are fallow swidden forest or tree-crop gardens which are under the customary ownership of individual villagers. These lands are called *Krat* in the local language (Table 11). From this we conclude that villagers adopt “portfolio land-use strategies” when confronted with oil palm plantation development[23].

Table 11. Villagers’ intentions for oil palm plantation development (n = 14)

Refuse in Bengkar; Refuse in Krat	2
Accept in Bengkar; Refuse in Krat	11
Refuse in Bengkar; Accept in Krat	0
Accept in Bengkar; Accept in Krat	0
No answer	1

Source: Based on Terauchi et al., (2010) and Terauchi (2013).

Criteria and Factors for Villagers’ Land-use Decision-Making and Land-Use Strategies

One can find three criteria for land-use decision-making based on villagers’ perceptions of rattan, rubber, and oil palm cultivation (Figure 2).

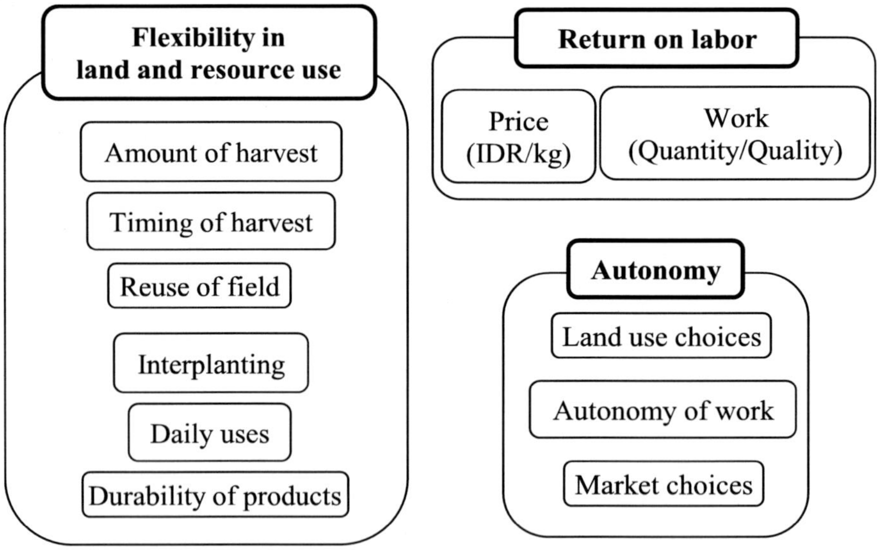

Figure 2. Criteria and factors of land-use decision-making.

Source: Modified from Terauchi et al., (2010) and Terauchi (2013).

[23] The definition of “portfolio” in finance is a collection of investments. “It is a generally accepted principle that a portfolio is designed according to the investor’s risk tolerance, time frame and investment objectives… When determining a proper asset allocation one aims at maximizing the expected return and minimizing the risk (Wikipedia, 2015)”. Actually, it is not proper to express villagers’ land-use strategy (investment) regarding oil palm cultivation as a “portfolio” because the *Krat* is an individual asset and the *Bengar* is a communal asset. In other words, this is not an asset-allocation strategy of each individual villager. But we use the term “portfolio land-use strategy” to describe villagers’ land-use strategies in response to oil palm plantation development for the purpose of maximizing the expected return and minimizing the risk.

The first is return on labor. Factors of this criterion are the price of tree crops and labor. Labor includes qualitative aspects (characteristics of labor) and a quantitative aspect (labor time). The importance of the qualitative aspect of labor was revealed through villagers' evaluation of rattan and rubber cultivation, as described above.

The second is "flexibility of land and resource use." For example, villagers note that one advantage of rattan is being able to harvest the needed amount at any time. Flexibility of harvest amount and time is an important factor in villagers' evaluation of land use. Also, the importance of repurposing gardens for swiddens, conversion to other gardens, and interplanting in gardens came to be understood by comparing rattan and rubber cultivation with oil palm cultivation. In the case of rattan, its many practical uses are seen as an advantage, while the advantage of rubber is that, because it keeps well, villagers can store it when prices are low, and sell it when prices rise.

Table 12. Perceptions of villagers in Besiq Village toward rattan, rubber, and oil palm cultivation[1]

		Rattan	Rubber	Oil palm[2]
Return on labor	Price	IDR 1300/kg	IDR 11,200./kg	IDR 1353/kg
	Labor (Quality, Quantity)	Management: + Harvesting: - Transport: -	Management: - Harvesting: + Transport: +	Management: (-) Harvesting: (+) Transport: n/a
Flexibility in land and resource use		Amount of harvest: + Timing of harvest: + Reuse of field: + Interplanting: + Practical uses: + Durability of product: -	Amount of harvest: - Timing of harvest: + Reuse of field: + Interplanting: + Practical uses: - Durability of products: +	Amount of harvest: n/a Timing of harvest: n/a Reuse of field: (-) Interplanting: (-) Practical uses: (-) Durability of products: (-)
Autonomy		Land use choices: + Work autonomy: + Market choices: +	Land use choices: + Work autonomy: + Market choices: +	Land use choices: (-) Work autonomy: (-) Market choices: (-)

Note 1: + and - mean positive and negative evaluations by local people, respectively.

Note 2: Parenthetical oil palm cultivation evaluations are not based on actual cultivation experience but on information from people in other villages and on explanatory meetings held by companies. "n/a" is used for transport labor, amount of harvest, and timing of harvest because local people did not mention them.

Source: Modified from Terauchi (2010), Terauchi et al., (2010), and Terauchi (2013).

The third is autonomy in choosing the type of land use, market choice, and type of work. Villagers had concerns that land use-choices such as land use conversion and interplanting in tree crop gardens are constrained by collaboration with companies. Land use conversion and interplanting as related to autonomy are evaluated in terms of social aspects. On the other hand, repurposing of fields and interplanting for the flexible use of land and resources are evaluated from the aspect of land use and cultivation technique. For example, villagers could actually repurpose Plasma fields as swidden fields and convert Plasma fields to other gardens, although according to villagers it is technically difficult. This example is a matter of flexibility of land and resource use. But conversion of Plasma fields to other gardens is prohibited by contracts with companies. This is a matter of autonomy. The importance of autonomy in choosing markets and type of work came to be understood in light of the villagers' concerns about collaboration with oil palm plantation companies. Villagers were afraid of the single-buyer FFB market and forced or heavy labor for companies.

Table 12 summarizes villagers' perceptions of rattan, rubber, and oil palm cultivation based on the criteria of land-use decision-making. Rattan cultivation is seen as having a low return on labor but high flexibility of land and resource use. Therefore villagers will not convert rattan gardens to other uses, but maintain them for practical uses and as a cash income source for emergencies. Rubber cultivation is seen as having a high return on labor. Villagers therefore increase rubber garden acreage for a regular cash income source. Concerns about oil palm cultivation are the low autonomy and low flexibility of land and resource use, but villagers have high expectations that the high return on labor will help improve their household economies. Hence villagers accept oil palm plantation development only in the customary forest of the upper river area.

This leads to the conclusion that villagers combine rattan and rubber cultivation because they complement one another (hybrid land-use strategy), and accept oil palm cultivation while partially avoiding social concerns (portfolio land-use strategy). We call this land-use combination the "mosaic land-use strategy."

Problems of Oil Palm Cultivation from the Perspective of Swiddeners

Here we discuss the problems of oil palm cultivation based on the realities of oil palm cultivation in Paser District, in conjunction with the three criteria of villagers' land-use decision-making in Besiq Village.

Return on Labor[24]

Some villagers in Paser District who have oil palm plantations achieved high standards of living, such as a high level of education for their children, luxurious homes, and ownership of a car and several motorbikes, thanks to high cash income from oil palm cultivation. Some villagers meet all their livelihood requirements with cash income from oil palm cultivation. Paser District villagers noted that the advantages of oil palm cultivation are the high cash income and the low harvesting labor requirement of once in two weeks. On the other hand, they gave rubber cultivation a low rating because of the high harvesting labor requirement of almost every morning (four or five days a week). Many villagers establish oil palm plantations without collaborating with companies or accepting any kind of government support. The hope of villagers in Besiq Village for a high return on labor is likely to be achieved in Paser District. Based on these findings, villages which, like Besiq Village, have enough land will accept oil palm cultivation to improve their standard of living somewhat.

It is important note, however, that the high return on labor of oil palm cultivation is realized after meeting prerequisites, namely leveling land and building roads for transporting FFBs. One must remember that the criteria for return on labor include labor characteristics as well as labor time. Companies pay villagers about IDR 1353/kg for FFBs, which is almost the

[24] This subsection is based on a modification of Terauchi (2013).

same as the price of rattan. Oil palm cultivation achieves high profitability because oil palms produce 15.7t/ha of FFBs, which means that villagers must transport a heavy product as they do with rattan. Rattan cultivation is regarded as yielding a low return on labor due to the dangerous work of carrying the heavy rattan despite the low price. As such, villagers hesitate to harvest rattan as a regular cash income source. It is possible that oil palm cultivation is also seen as offering a low return on labor. Moreover, transporting FFBs is more physically demanding and difficult than carrying rattan because it is difficult to remove thorns from FFBs, which are not as sharp as rattan thorns, and to bundle FFBs like rattan to carry them easily. Villagers in Paser District did not feel that transport is physically demanding and difficult because the company leveled the land and built roads for transporting FFBs. Villagers can carry FFBs by handcart. Hence oil palm cultivation can achieve a high return on labor only after high capital investment to level land and build roads. Especially in inland places such as West Kutai District, where land is not level and there are few roads, the return on labor of oil palm cultivation was considered to be almost the same as that of rattan cultivation and less than that of rubber cultivation. Although villagers in Besiq Village are willing to cultivate oil palms by themselves under current conditions, they will likely be disappointed with the low return on labor of oil palm cultivation.

Researchers ordinarily evaluate agricultural productivity in terms of labor productivity and land productivity. It appears that companies and the government promote oil palm cultivation based on high land and labor productivity. But in a context where land is still plentiful but labor is scarce, labor productivity is more important than land productivity for villagers in land-use decision-making (Feintrenie et al., 2010a). Although we concur with other researchers on this point, our study found that villagers evaluate land use in terms of their return on labor rather than labor productivity. It is important to close the gap separating perceptions of profitability between villagers on the one hand, and companies and the government on the other, and to incorporate villagers' views into development projects.

Flexibility of Land and Resource Use

Companies establish oil palm plantations only to harvest FFBs, but Paser District villagers establish their own plantations in various ways and for multiple purposes. In the first and second years, most villagers grow rice and vegetables in addition to planting oil palms. Some villagers plant oil palms in existing rubber gardens or banana gardens (Photo 4). One villager manages 25 useful plants such as fruit trees and medicinal plants in one oil palm field (Terauchi, 2011a). It is an important finding that Paser District villagers also prefer flexible land and resource use even though villagers think that FFBs and tree crops interplanted in oil palm fields have poor productivity. In addition, villagers collect, eat, and sell many mushrooms, which are raised from pruned palm fronds and discarded empty fruit bunches carried from mills to Inti fields (Terauchi, 2011b).

In 2002, the Forestry Service and Plantation Service of Paser District collaborated and implemented a 255-ha oil palm agroforestry program, which is one of choices in the Forest and Land Rehabilitation (*Rehabilitasi Hutan dan Lahan*) program. Participating villagers prepare 1 ha of land to plant oil palm trees and teak trees. Harvesting FFBs was to start in the fourth year, and teak was to be harvested in the 12th year for sale or for use by villagers as construction material, but teak did not grow well and its cultivation failed because teak loses in competition with oil palms. The Paser District Forestry Service and Plantation Service admitted the failure of the agroforestry program combining oil palms and teak (Terauchi, 2011b).

In 2006 the Paser District Livestock Service implemented a program which integrates cattle rising and oil palm plantations. Villagers participating in this silvopastoral program lease a few head of cattle from the service and breed them. Instead of returning the same number of calves to the service, villagers give them to other participating villagers. The purpose of this program is not only to create another cash income source in addition to oil palm cultivation, but also to ensure that villagers will have cattle to transport FFBs and to eat pruned palm fronds. In Paser District, villagers freely graze cattle in oil palm plantations. Cattle eat the weeds and fertilize plantations with their manure. Sometimes cattle dung can be sold at IDR 4000/kg to collectors who sell it to other farmers as plantation fertilizer. Villagers say that it costs little to take care of cattle. According to the Livestock Service, cattle sold for IDR 27,000/kg and beef sold for IDR 80,000/kg in 2009 (Terauchi, 2011b). It appears that the silvopastoral practice of oil palm cultivation and raising cattle is highly synergetic.

Examining oil palm cultivation in Paser District reveals different forms of plantation management, such as the inter-planting of useful plants, collecting mushrooms, agroforestry, and silvopasture based on oil palm cultivation, even though it is said that oil palm plantations are monocultures. These practices and trials are in line with the villagers' criterion of flexibility of land and resource use. It is important that the preference for such flexibility contributes objectively (that is, not in the villagers' subjective view) to avoiding market-economy risks, although villagers in Besiq and in Paser District also want to ensure such flexibility in order to meet their daily needs.

Photo 4. Rubber garden interplanted with oil palm (Source: Terauchi (2011a)).

Another way of ensuring flexibility of land and resource use is appropriate land-use zoning that divides land for oil palm plantation development from land that is conserved for customary forest and land use. This division is in line with the portfolio land-use strategy of Besiq villagers. Since the end of the Suharto regime, indigenous people can choose whether to accept oil palm plantation development. In other words, villagers decide what is appropriate land-use zoning. But there is a structural political power imbalance between villagers and companies: Villagers can obtain information only through companies, but companies can choose to provide villagers only with information favorable to the companies (Terauchi, 2013). To eliminate this imbalance, it is necessary to ensure and support free, prior, and informed consent (FPP, 2015). That is to say, full information about proposals should be provided beforehand, and villagers' decisions should not be made under duress or skewed by corruption (Marti, 2008). Villagers must be able to participate in making decisions on land-use zoning. Multiple actors such as governments, NGOs, and researchers need to participate in the process of land-use zoning and support appropriate decision-making by villagers.

Autonomy

Villagers in Besiq Village had concerns that if they accepted oil palm plantation development, contracts with companies would limit their land-use choices, such as by banning interplanting in Plasma fields and prohibiting conversion of Plasma fields to other gardens. In Paser District, Plasma fields made a quite high economic contribution to household economies, and therefore almost no one wants to convert Plasma fields to other tree crop gardens. And in reality, villagers who have not yet paid back Plasma loans are not allowed to convert from Plasma to other land use. But if villagers obtain title to their land after finishing repayment, they can use the land freely. Interplanting fruit trees and useful plants in oil palm fields was conducted not in Plasma fields but in oil palm fields established by villagers themselves.

Regarding work autonomy, villagers in Besiq Village are concerned that companies will enforce harsh working conditions in Plasma fields, but the reality in Paser District was different. Villagers who have Plasma fields manage and harvest them autonomously. Most villagers do not use adequate amount of fertilizers or herbicides, whose use is subscribed by company guidelines, because they lack the cash to buy them (Terauchi, 2011a). Besiq villagers are also worried about harsh working conditions in Inti fields. NGOs reported that some companies set worker quotas for harvest amounts and field management operations in Inti fields, and if workers miss their quotas, their incomes are reduced by corresponding amounts. Herbicides that are banned in many countries are used freely; they cause serious human health problems and have even lead to death (Touge, 2002). However, for paid Inti labor, the national company in Paser District did not set unachievable quotas. Harvest and management workers can achieve their quotas in the working period from 7:00 to 12:00 (Terauchi, 2011a). Harvest workers sometimes continue to harvest until 15:00 to obtain additional income. Owing to democratization in post-Suharto Indonesia, villagers have started mounting protests against the adverse decisions and practices of companies. In Paser District the national company has been considerate of local people. But there is little information about private companies because it is difficult to research them.

Regarding market choices, Besiq villagers are concerned about restrictions on how they sell their FFBs. Villagers fear boycotting or unjustly low FFB prices by companies in a single-buyer market. But it appears there is no high interdependence between villagers and contracted oil palm companies in Paser District. Several oil palm companies' mills and many FFB brokers have emerged in Paser District. Even some villagers have become FFB middleman collectors between villagers and broker companies, thereby creating competition for villagers' FFBs. Some villagers who have Plasma fields ship FFBs to broker companies because of the high prices they can get, instead of to their contracted companies' mills. Villagers prefer FFB collectors other than a contracted company's mill, even if that means a lower FFB price, because villagers can harvest FFBs and obtain cash income whenever they want. Generally, villagers call a collector and say that they will harvest FFBs that day. The FFB collector buys the FFBs and pays the villagers on the spot. If villagers ship FFBs to company mills, they must wait until a particular day for payment, which is once a month.

CONCLUSION

This chapter describes swiddeners' perceptions of land use and land-use strategies in a case study of Besiq Village, West Kutai District, which is the frontier of oil palm plantation development in East Kalimantan. Our study found that people in Besiq Village evaluated land uses such as rattan, rubber and oil palm cultivation according to three criteria: return on labor, flexibility of land and resource use, and autonomy. Rubber cultivation was chosen as a source of cash income because of its high return on labor. Rattan cultivation is maintained as a source of cash income in emergencies, and to obtain materials for practical utensils, because of the high flexibility of rattan cultivation in land and resource use despite its low return on labor. Villagers hesitated to introduce oil palm cultivation because of its low flexibility of land and resource use and restriction on autonomy, although they had expectations for its high return on labor. Therefore villagers accepted it in their customary forest of the upper river area but refused it in the vicinity of the village, where the lands are de facto owned by villagers. We call this land-use combination a "mosaic land-use strategy," wherein villagers combine rattan and rubber cultivation because they are complementary (hybrid land-use strategy), and accept oil palm cultivation while partially avoiding social concerns (portfolio land-use strategy).

We also checked whether oil palm cultivation fulfills the three evaluation criteria by examining a case from Paser District, which is on the frontier of the palm oil industry in East Kalimantan. We found that villagers' expectations for and concerns about oil palm cultivation in Besiq are much different than the reality of oil palm cultivation in Paser District.

Villagers in Besiq had high expectations for the return on labor of oil palm cultivation. Villagers in Paser District indeed evaluated oil palm cultivation as having a high return on labor, and they achieved a high standard of living, but there are important prerequisites. If land is not leveled and roads are not built sufficiently as in West Kutai District, the return on labor of oil palm cultivation is considered to be almost the same as that of rattan cultivation, and less than that of rubber cultivation.

Villagers in Besiq had concerns about the low flexibility of land and resource use by oil palm cultivation. Although it is said that oil palm plantations are pure monocultures, oil palm cultivation in Paser District shows that villagers make plantations more diverse by practicing agroforestry-based oil palm cultivation. They produce rice and vegetables in the land-clearing stage of oil palm plantations, and they plant and manage oil palms with rubber and banana trees and other useful plants in the same fields. They also collect mushrooms in company-owned plantations. In addition, the local government implemented agroforestry and silvopastoral programs for villagers based on oil palm cultivation. Although some of these practices have not yet realized good harvests, it is possible that flexibility of land and resource use has been improved even in oil palm cultivation. It is also effective to establish land-use plans through a participatory approach in which certain forest zones are not converted to oil palm plantations, but instead reserved for the use of local people.

Besiq villagers also had concerns about the low autonomy of oil palm cultivation, but we observed no serious problems with autonomy in Paser District. In particular, autonomy in work and product sales improved after democratization in 1998; freedom in this respect has been promoted and has created a seller's market for FFBs. The national oil palm company in Paser District started paying attention to the desires of local people after democratization, although it is not clear that private companies have followed suit. The establishment of mills by several companies and the emergence of middlemen who collect FFBs gave villagers a choice of FFB buyers.

Companies and the government promote large-scale oil palm plantation development despite the perception gap and information disparity regarding oil palm cultivation between local people on one hand, and companies and the government on the other. The government and companies must understand local peoples' perceptions of land use — criteria such as return on labor, flexibility of land and resource use, and autonomy — and modify development plans in accordance with local peoples' hopes and concerns. Local people must also be provided with appropriate information about oil palm cultivation. There are big differences between the expectations and concerns of villagers in Besiq, which is the frontier of oil palm plantation development, and the realities of oil palm cultivation in Paser District, which is on the frontier of the palm oil industry in East Kalimantan. It is likely that not only Besiq Village, but also other villages on the frontier of oil palm plantation development, find themselves in the same situation. It is important that third parties such as NGOs and researchers actively work to align expectations with reality.

Future Research

This study elucidated swiddeners' perceptions of major land uses, i.e., rattan, rubber, and oil palm cultivation, and land-use strategy. In future research, we need to determine swiddeners' perceptions of livelihood activities including not only rattan, rubber, and oil palm cultivation, but also other activities and overall livelihood strategies. In addition, we explained the current situation with oil palm cultivation in Paser District. But future research must determine the historical change in the palm oil industry in Paser District as it corresponds with historical change in the macro socioeconomic structure. This knowledge is important for examining the possibilities and risks of the palm oil industry.

REFERENCES

BPS (Badan Pusat Statistik). (2013). *Kutai Barat dalam angka (West Kutai in figures) 2007.* Babupaten Kutai Barat: Badan Pusat Statistik.

Chambers, R. (1983). *Rural development: Putting the last first.* London: Longman.

Chambers, R. (1997). *Whose reality counts?: Putting the fast last.* London: Intermediate Technology.

Cramb, R. A. (1993). Shifting cultivation and sustainable agriculture in East Malaysia: A longitudinal case study. *Agricultural Systems*, 42(3), 209-226.

Cramb, R. A. & Sujang, P. S. (2013). The mouse deer and the crocodile: Oil palm smallholders and livelihood strategies in Sarawak, Malaysia. *The Journal of Peasant Studies*, 40(1), 129-154.

de Jong, W. (2001). The impact of rubber on the forest landscape in Borneo. In A. Angelsen, D. Kaimowitz (Ed.), *Agricultural technologies and tropical deforestation* (pp. 367-381). Wallingford, UK: CABI.

DK (Departemen Kehutanan). (2008). *Statistik kehutanan Indonesia tahun 2008.* Jakarta: Departemen Kehutanan.

Dove, M. R. (1993). Smallholder rubber and swidden agriculture in Borneo: A sustainable adaptation to the ecology and economy of the tropical forest. *Economic Botany*, 47, 136-147.

DPPKT (Dinas Perkebunan Provinsi Kalimantan Timur). (2013). *Statistik perkebunan Kalimantan Timur 2012.* Provinsi Kalimantan Timur: Dinas Perkebunan.

FAO (Food and Agriculture Organization). (2013, April 29). FAOSTAT online statistical service. Retrieved from http://faostat3.fao.org/home/index.html#DOWNLOAD.

Feintrenie, L., Chong, W. K., & Levang, P. (2010a). Why do farmers prefer oil palm? Lessons learnt from Bungo district, Indonesia. *Small-scale Forestry*, 9(3), 379-396.

Feintrenie, L., Schwarze, S., & Levang, P. (2010b). Are local people conservationists? Analysis of transition dynamics from agroforests to monoculture plantations in Indonesia. *Ecology and Society*, 15(4), 37.

FPP (Forest Peoples Program). (2015, May 5). Free, prior and informed consent (FPIC). Retrieved from http://www.forestpeoples.org/guiding-principles/free-prior-and-informed-consent-fpic.

Godoy, R. A. (1990). The economics of traditional rattan cultivation. *Agroforestry Systems*, 12, 163-172.

Gönner, C. (2011). Surfing on waves of opportunities: Resource use dynamics in a Dayak Benuaq community in East Kalimantan, Indonesia. *Society & Natural Resources*, 24(2), 165-173.

Gönner, C., Cahyat, A., Haug, M., & Limberg, G. (2007). *Towards wellbeing: Monitoring poverty in Kutai Barat, Indonesia.* Bogor, Indonesia: Center for International Forestry Research.

Gönner, C. & Seeland, K. (2002). A close-to-nature forest economy adapted to a wider world. *Journal of Sustainable Forestry*, 15(4), 1-26.

Gouyon, A., de Foresta, H., & Levang, P. (1993). Does 'jungle rubber' deserve its name? An analysis of rubber agroforestry systems in southeast Sumatra. *Agroforestry Systems*, 22(3), 181-206.

Inoue, M. (1989). *Final report: Swidden cultivation in East Kalimantan*. The Tropical Rain Forest Research Project JTA 9 (a)-137 Japan International Cooperation Agency.

Inoue, M. (1991). The system of swidden and rattan production in East Kalimantan, Indonesia. *Journal of Forest Economics*, 119, 43-50.

Inoue, M. & Lahjie, A. M. (1990). Dynamics of swidden agriculture in East Kalimantan. *Agroforestry System*, 12, 269-284.

Kawai, M. (2011). "Moderate industrialization" and commons: Alternative development strategy to oil palm plantation in East Kalimantan, Indonesia. 13th Biennial Conference of the International Association for the Study of the Commons (IASC), 12th January 2011, Hyderabad, India.

Kawai, M. & Inoue, M. (2010). Possibility of moderate industrialization as alternative development strategy to oil palm plantation: The case of middle-upper Mahakam region in East Kalimantan, Indonesia. *Forest Economy*, 63(7), 1-17.

King, T. (1993). The peoples of Borneo. Oxford: Blackwell.

Marti, S. (2008). *Losing ground: The human rights impacts of oil palm plantation expansion in Indonesia*. Friends of Earth, Life Mosaic, & Sawit Watch Indonesia.

Pahan, I. (2010). *Panduan lengkap kelapa sawit: Manajemen agribisinis dari hulu hingga hilir*. Jakarta: Penebar Swadaya.

Pambudhi, F., Belcher, B., Levang, P., & Dewi, S. (2004). Rattan (*Calamus spp.*) gardens of Kalimantan: Resilience and evolution in a managed non-timber forest product system. In K. Kusters. & B. Brian. (Ed.), *Forest Products, Livelihoods and Conservation: Case Studies of Non- Timber Forest Product Systems Volume 1-Asia* (pp. 337-354). Bogor, Indonesia: Center for International Forestry Research.

Pelzer, K.J. (1978). Swidden cultivation in Southeast Asia: Historical, ecological, and economic perspectives. In P. Kunstadter, E. C. Chapman, & S. Sabhasri. (Ed), *Farmers in the forest: Economic development and marginal agriculture in northern Thailand* (pp. 271-286). Honolulu: University press of Hawaii.

Penot, E. (2004). From shifting agriculture to sustainable complex rubber agroforestry systems (jungle rubber) in Indonesia: A history of innovation processes. In D. Babin (Ed.), *Beyond Tropical Deforestation from tropical deforestation to forest cover dynamics and forest development* (pp. 221-249). Paris: The United Nations Educational, Scientific and Cultural Organization and Centre de coopération internationale en recherche agronomique pour le développement.

Rist, L., Feintrenie, L., & Levang, P. (2010). The livelihood impacts of oil palm: Smallholders in Indonesia. *Biodiversity and Conservation*, 19(4), 1009-1024.

Sheil, D., Casson, A., Meijaard, E., van Noordwijk, M., Gaskell, J., Sunderland-Groves, J., Wertz, K., & Kanninen, M. (2009). *The impacts and opportunities of oil palm in Southeast Asia, what do we know and what we need to know?* Occasional Paper No. 51. Bogor, Indonesia: Center for International Forest Research.

Terauchi, D. (2010). Livelihood strategies of Borneo's swiddeners: Tree crop changes from rattan to rubber, and eventually to oil palm? In G. Mitsumata, Y. Suga & M. Inoue (Ed.), *The Possibility of the Local Commons: New Relationship between People and Environment towards Autonomous Governance* (pp. 89-114). Kyoto: Minerva Shobou.

Terauchi, D. (2011a). The frontier of oil palm production in East Kalimantan (1). *Japanese Journal of International Forest and Forestry*, 80, 41-46.

Terauchi, D. (2011b). The frontier of oil palm production in East Kalimantan (2). *Japanese Journal of International Forest and Forestry*, 81, 36-41.

Terauchi, D. (2013). The struggles of Borneo's swiddeners in the era of oil palm plantation development. In M. Koike & A. Miki (Ed.), *Forest and Silviculture Works in Japan and Asia* (pp. 193-226). Nagano: Kawabe Shorin.

Terauchi, D. & Inoue, M. (2011). Changes in cultural ecosystems of a swidden society caused by the introduction of rubber plantation. *Tropics*, 19, 67-83.

Terauchi, D., Setsuda, T., & Inoue, M. (2010). Preference of swiddeners to rattan, rubber, and oil palm: Based on the evidence from Besiq Village, Indonesia's province of East Kalimantan. *Journal of the Japanese Forest Society*, 92, 247-254.

Touge, R. (2002). What's happened in oil palm plantation in Malaysia?: Plantation workers suffering from pesticide pollutions. In S. Okamoto (Ed.), *Light and shadow of oil palm plantation development: What's happened in Indonesia and Malaysia* (pp. 22-26). Japan: Japan NGO Network on Indonesia.

Weinstock, J. A. (1983). Rattan: Ecological balance in a Borneo Rainforest. *Economic Botany*, 37, 58-68.

Wikipedia. (2015, August 8). Portfolio (finance). Retrieved from https://en.wikipedia.org/wiki/Portfolio_(finance).

Zen, Z., Barlow, C., & Gondowarsito, R. (2006). Oil palm in Indonesian socio-economic improvement: A review of options. *Oil Palm Industry Economic Journal*, 6(1), 18-29.

In: Monoculture Farming
Editors: Tapan Kumar Nath and Patrick O'Reilly
ISBN: 978-1-63485-166-4

Chapter 6

ALTERNATIVE DEVELOPMENT STRATEGIES FOR LARGE SCALE OIL PALM PLANTATIONS IN EAST KALIMANTAN, INDONESIA

***Masayuki Kawai*[1,*] *and Makoto Inoue*[2]**
[1]Institute for Global Environmental Strategies
[2]The University of Tokyo, Japan

ABSTRACT

Oil palm plantations are rapidly expanding along with negative environmental and social effects, such as irreversible large-scale deforestation, biodiversity loss and wide-ranging land disputes in Indonesia. This chapter addresses some of the aforementioned problems by considering alternative development strategies for large-scale oil palm plantations, focusing on the smallholder development policies and institutions in Indonesia. The Nucleus Estate and Smallholders (NES) and Project Management Unit (PMU) schemes have existed since the 1970s during the era of Suharto's authoritarian regime. The former was supported by estate companies and mainly applied to oil palm plantations, while the latter was supported by government project management units and mainly applied to rubber and coconut plantations. The important aspect here is that the PMU scheme can avoid rapid and large-scale land expropriation by estate companies, which often becomes a problem in the NES scheme because the PMU scheme supports smallholders without the involvement of the companies, meaning that 'well thought-out' and 'unhurried' development, which is known as 'moderate industrialization,' can be considered to be one of the alternative development options. This moderate industrialization might be realized by establishing small, dispersed but modern high-yield smallholdings through the PMU scheme that at the same time maintains traditional forestry practices and diverse local livelihoods, such as swidden, orchard and rubber agroforest maintenance. The PMU scheme applied at the national level, however, is facing difficulties that stem from its institutional shortfalls, such as high input/investment and the lack of an effective credit repayment system and land tenure uncertainty in local communities. This chapter suggests some possibilities based on this case study to cope

* Corresponding Author address2108-11 Kamiyamaguchi, Hayama, Kanagawa, 240-0115 Japan. Email: kawaimo76@gmail.com.

with these problems through a more flexible use of the PMU scheme. There is an urgent need to develop economically attractive but ecologically and socially sound smallholder support institutions that can alleviate the strong expansion pressure of companies that lead to large-scale oil palm monoculture and marginalization of local people through land expropriation.

Keywords: oil palm, NES, PMU, smallholder, Indonesia

INTRODUCTION

The presence of the oil palm (*Elaeis guineensis*) is rapidly expanding, causing deforestation and a loss of biodiversity in Indonesia. The oil palm area increased from 1.13 million ha in 1990 to 10.47 million ha in 2013 according to the data from the statistics of Ministry of Agriculture. The oil palm is well known as one of the most profitable oil crops, and palm oil has many uses — for example, in foods, detergents, cosmetics, and surfactants (Douglas et al., 2009). Interest in the oil palm for biofuel use has been increasing (ibid). However, the profitable and useful characteristics of palm oil encounter difficulties in terms of alleviating the conversion of tropical forest areas into oil palm plantations (Willcove and Koh 2010). Between 1990 and 2005, the minimum forest area that was converted to oil palm plantations was estimated to be 1.7 million ha in Indonesia (Koh and Wilcove 2008). In Kalimantan, an estimated 90% of land converted to oil palm from 1990 to 2010 was forest (47% intact, 22% logged, 21% agroforest) (Carlson et al., 2013).

In terms of social impact, many land disputes and conflicts between oil palm companies and local people were reported by NGOs (Marti 2008). After Suharto resigned in 1998, however, local people's political presence increased, and many companies experienced resistance in the form of demonstrations and were required to compensate the local people to improve their socio-economic conditions. Smallholders accelerated oil palm expansion, and the number of studies highlighting the profitability of oil palm for smallholders' livelihoods is increasing (Zen et al., 2006, Rist et al., 2010, Feintrenie et al., 2010), even though many social, environmental and political challenges related to rapid agrarian changes led by oil palm development remain, such as large-scale traditional land/forest claims by companies, uneven participation and political power imbalances between smallholders and companies that are related to local patronage politics and governance, which are diverse and dependent in each local context (McCarthy 2010, Gillespie 2012, McCarthy et al., 2012) Coping with and mitigating such negative contradictions resulting from oil palm development require urgent attention.

This chapter addresses part of the aforementioned problems by considering alternative development options for large-scale oil palm plantations. In the era of Suharto's authoritarian regime, there were two large smallholder development schemes — that is, the Nucleus Estate and Smallholders (NES) and Project Management Unit (PMU) schemes. The former was supported by estate companies while the latter was supported by government project management units, not estate companies. These two schemes are known as integrated schemes that provide high-yielding clone seedlings, fertilisers, agrichemicals, farming tools, technical advice, land tenure and securing markets and contribute to the development of smallholders. The framework then passed to an estate revitalization program implemented during the democratized era (2006-2014).

This chapter clarifies the impact of these two schemes on local peoples' socio-economic conditions in East Kalimantan and considers the possibilities and challenges of alternative development strategies for large-scale oil palm plantations through the modernization of smallholders in the current socio-economic circumstances in Indonesia by comparing these two schemes.

METHODS

The impact of the NES and PMU schemes was clarified through a literature review and field research.

The Overview of NES and PMU Schemes at the National Level

The overview of the development of the NES and PMU schemes and the outreach at the national level are clarified by literature reviews that include academic articles (Barlow and Jayasuriya 1984; Barlow and Tomichi 1991; Zen et al., 2006), an NGO report (Marti 2008), the project planning or implementation reports of the Ministry of Agriculture of Indonesia (Departemen Pertanian 1984; Departemen Pertanian 1985; Departemen Pertanian 1986; Departemen Pertanian 1992; Departemen Pertanian 1999; Departemen Pertanian 2006; Departemen Pertanian 2007), the project evaluation report of the Ministry of Forestry and Estate Crops of Indonesia (2000), and the project completion report of Asian development banks (Asian Development Bank 1997; Asian Development Bank 2002; Asian Development Bank 2004a; Asian Development Bank 2004b; Asian Development Bank 2005).

The Impact of NES and PMU Schemes at the Local Level in East Kalimantan

The impact of NES and PMU schemes at the local level in East Kalimantan is clarified by using the authors' field research data, collected mainly during the 2007-2009 period.

Research Site

Figure 1 shows the research sites in Mahakam Ulu[25], the West Kutai district and the Paser district.

The Middle-Upper Mahakam region was selected as the case study site for the pre-development area. This region consists of Laham, the Long Hubung sub-district in the Mahakam Ulu district and the Long Iram sub-district in the West Kutai District. The population was 18,503, the area was 289,471 ha, and the population density was 6.39 per capita per km^2 as of 2008. The main ethnic group is the Bahau Dayak people. Rich secondary

[25] Mahakam Ulu district was established in 2013. At the research time the current area of Mahakam Ulu district was part of West Kutai district.

forest area has remained in this region, but the primary forest area was mostly destroyed by commercial logging and forest fires. The main transportation methods consisted of small boats with an outboard engine that travelled on the Mahakam River and its branches. Most villages were located along these rivers. Only unpaved logging roads existed, and it was difficult to access this region via land routes.

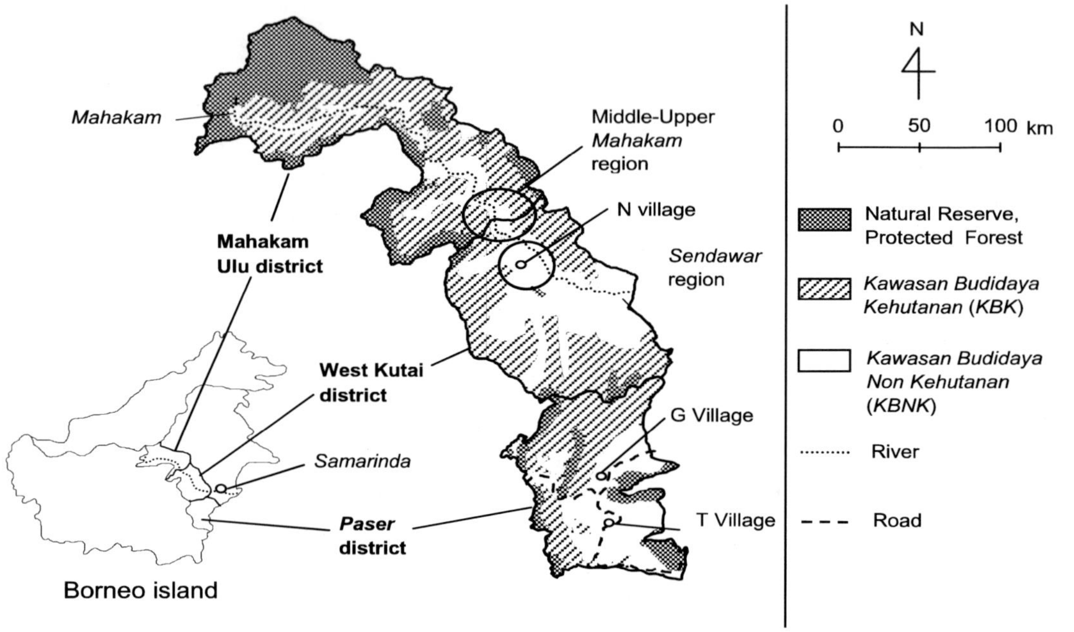

Figure 1. Research sites in Mahakam Ulu[26], the West Kutai district and the Paser district.
Note. The Natural Reserve, Protected Forest, and *Kawasan Budidaya Kehutanan* (*KBK*) are categorised as state-owned forests (*Kawasan Hutan*), and these land are under the control of the Ministry of the Environment and Forestry. In contrast, *Kawasan Budidaya Non Kehutanan* (*KBNK*) is categorised as other use land (*Areal Pengunaan Lain*) and is managed by the local government. *KBNK* is designated as multipurpose use, such as for cities, towns, villages, and farmland for swidden, fruit and other plantations. Private land tenure is approved only in *KBNK*. Notably, the establishment of oil palm plantations is allowed only in *KBNK*.

The Paser district was selected as the case study for oil palm plantations. It is the first district since 1983 in East Kalimantan in which an oil palm plantation was developed. The Paser district consists of 10 sub-districts and 117 villages. In 2008, the population was 191,117, the area was 1,160,396 ha, the population density was 16.47 km^2, and the total area of the oil palm plantation in the Paser district was 95,822 ha. The Paser people are indigenous to the Paser district. Because most Paser people are Muslim, they are distinguished from Dayaks, a general term for non-Muslim indigenous people in Borneo. However, the people of Paser have many similarities with the Dayak people because both have lived with slash-and-burn agriculture. T village and G village were selected as the research sites in the Paser district. Both villages have nucleus estates and oil palm mills established by state-owned companies. The villages are located at the centre of the oil palm industry.

[26] Mahakam Ulu district was established in 2013. At the research time the current area of Mahakam Ulu district was part of West Kutai district.

As the case study for rubber smallholding established by the PMU scheme, the Sendawar region was selected, which consists of the Melak, Barong Tongkok, Linggang Bigung and Sekolak Darat sub-districts in the West Kutai District in this paper. The population was 50,678, the area was 164,484 ha, and the population density was 30.8 per capita per km^2 in 2007. From 1981 to 2001, 10,387 ha of high-yielding rubber smallholdings were developed by the PMU scheme in this region. The ethnic majority is the Tonyoi Dayak people. N village, one of the villages supported by the PMU scheme in this region, was selected as a research site.

Data Collection

Nine villages were visited in the Middle-Upper Mahakam region from 2007 to 2009. T village and G village in the Paser district were visited from July to August 2009. N village in the West Kutai district was visited in November 2008 and March 2009. The key informant interviews with village heads, traditional law heads, and villagers were conducted in each village, and the key informant interviews with government officers in the local estate office in the West Kutai district and Paser district were conducted from 2007 to 2009. The key informant interviews with the staff of the state-owned estate company (*Perseroan Terbatas Nusantara XIII, PTPN XIII*) in the Paser district were conducted in July and August 2009.

A structural household survey was also conducted to estimate the amounts of cash and non-cash income from oil palm plantations, rubber smallholdings, traditional farms and other subsistence activities. Accurate fixed income data, such as salaries from companies and public servants, were collected. Non-fixed income data, such as income from agriculture, trade and small business, however, were estimated by asking about the values of commodities and frequency of sales of these commodities. Conflicts in the data were corrected through re-interviews where possible. Interviews were conducted by the author (Kawai) in the Indonesian language.

In the Middle-Upper Mahakam region, a structural household survey was conducted with members of *Credit Union Petemai Urip* (*CUPU*)[27] in nine villages from November 2007 to March 2008. One hundred fifty members were selected randomly from 2,246 *CUPU* members, omitting members who were under 17 years old and staff and board members. Next, 101 households who agreed to participate were interviewed. In this paper, data from 48 households in the Middle-Upper Mahakam region are used for analysis.

In T village, a household survey was conducted in July 2009. Thirty-one households were interviewed. Data from 25 indigenous people's households are used for this paper. The sample households were selected by the village head of T village and some *RT* (*Rukun Tetangga*, a type of neighbourhood association) heads on a voluntary basis. Seventeen households had oil palm smallholdings including oil palm smallholdings supported by NES scheme and eight households do not have oil palm smallholdings.

In G village, the household survey was conducted in August 2009. Only the indigenous households with local government involved in the flexible PMU type scheme involving oil palm smallholdings were selected by the village head and village staff on a voluntary basis. Next, 20 households were interviewed randomly.

[27] This CU has operated in the Middle-Upper *Mahakam* region since 2002. Because CU serves both poor and wealthy people, this paper assumes that the CU members represent the economic condition of the local people in the Middle-Upper *Mahakam* region, although there is a possibility that there is a bias in that the proportion of rich households who are *CUPU* members is slightly higher than that of non-*CUPU* members.

In N village in the West Kutai district, 20 sample households were randomly selected in *RT* 3[28], and household surveys were conducted in March 2009. Eighteen households have PMU scheme rubber smallholdings, one household has non-support rubber smallholdings and one household works on another person's PMU rubber smallholdings.

Because the sampling methods were different in each region and village, there is some bias, and the sample households could not represent all residents from the village or region. These household survey data, however, may be useful for estimating the condition of each village's or region's residents in a certain range by performing a cross-check with qualitative data from formal and informal interviews.

The sample households are categorised based on their economic status as follows: Very Low: under US $0.60[29] per capita per day (a poverty line defined by the Indonesian statistics office in 2007); Low: more than US $0.60 to under $1.08 per capita per day (a poverty line defined by the World Bank); Middle: more than US $1.08 to under $2.00 per capita per day (an average poverty line in developing countries); and High: more than US $2.00.

Results and Discussion

Nucleus Estate and Smallholders (NES) Scheme

The first NES scheme project started in 1977 in the Aceh and South Sumatra provinces with the support of the World Bank for rubber (*hevea brasiliensis*) plantations. The government then applied for oil palm *(elaeis guineensis)*, coconut (*cocos nucifera*), tea (*camellia sinensis*) and sugar corn (*saccharum officinarum*) plantation support and expanded the NES scheme project to other provinces[30] (Departemen Pertanian 1992).

Characteristics of the NES Scheme

The main characteristics of the early NES scheme were

1) Integrating estate companies and smallholders and building reciprocal relationships between them to increase the productivity of smallholdings and simultaneously fulfil the demand by the mills for raw materials.
2) Implementation on 'undeveloped land' (including the primary forest and the local people's customary forest).
3) Establishing new economic spheres with a nucleus estate (*Inti*) of a company (in the case of oil palms, 3,000 ha - 20,000 ha of plantation is established per mill).
4) Accelerating employment absorption for transmigrants in combination with transmigration programs.

[28] There is no apparent socioeconomic difference in households between each *RT*.

[29] US $1 was equal to 9,140 Rp in 2007 and 9,666 Rp in 2008 as per Indonesian Central Bank annual reports.

[30] The NES scheme project was implemented in Aceh, North Sumatra, Riau, West Sumatra, Jambi, South Sumatra, Bengkulu, Lampung, West Jawa, Central Jawa, West Kalimantan, East Kalimantan, South Kalimantan, Central Kalimantan, Central Sulawesi, South Sulawesi and Irian jaya provinces (Departemen Pertanian 1999).

Concept of NES Scheme

A government report explained the background and concept of the NES scheme as follows.

'In Indonesia, a long plantation history — which has existed for more than 170 years — created two principal patterns with regard to plantation management, which was developed and later became two sub-systems: large estates and smallholder plantations. The long journey of these two patterns, along with their unique natural features, characteristics, speed and strength, have already brought out many discrepancies between the two. Occasionally the discrepancies are antagonistic but obviously dualistic (based on which Boeke later derived his famous theory[31]). However, why do these two sub-systems always contradict one other?

The NES concept does not ignore the existence of such dualism but identifies the most positive aspects of the existence of the two sub-systems. The NES concept tries to make the two systems into one in which there is interdependency. To that end, collaboration is necessary, which lasts only when the collaboration offers mutual benefit.

The creation of the new concept is not to generate a certain new theory but to respond to the most important challenge: how all the potential can be actuated to accelerate development in the estate sector. Because of this, the government bravely adopted this concept by starting the development of the first NES project in 1977 in two locations, Aceh and South Sumatra, with rubber. The World Bank met the government's strong will very positively, and support from the World Bank was realised in the form of financing support for this first NES project' (Departemen Pertanian 1992, Ringkas-1: translated by the authors).

'Sejarah panjang mengenai perkebunan di Indonesia yang telah berlangsung selama lebih kurang 170 tahun melahirkan dua pola utama dari segi pengelolaan perkebunan yang kemudian berkembang menjadi dua sub-systim, yaitu perkebunan besar dan perkebunan rakyat. Perjalanan panjang kedua pola tersebut sesuai dengan sifat, cirri, kecepatan dan kekuatan masing-masing telah memunculkan banyak diskrepansi diantara keduanya. Terkadang antagonistis, tetapi yang jelas dualistis (dari sinilah kemudian Boeke melahirkan teorinya yang tersohor itu). Tetati mengapa kedua sub-systim itu selalu dipertentangan?

Konsepsi Perkebunan Inti Rakyat tidak mengabaikan adanya dualisme tersebut, tetapi berusaha melihat sisi yang paling positip dari keberadaan dua sub-sistim itu. Konsepsi PIR mencoba meletakkan keduanya kedalam satu pigura dimana ada ketergantungan satu sama lain. Untuk itu perlu kerjasama. Dan hanya bisa langgeng apabila kerjasama itu saling menguntungkan.

Lahirnya konsepsi itu bukan untuk menyusun suatu teori baru, tetapi untuk menjawab tantangan yang paling pokok: bagaimana seluruh potensi yang diniliki dapat digerakkan untuk memacu pembangunan di bidang perkebunan. Karena itu dengan berani Pemerintah mewujud-nyatakan konsepsi tersebut dengan mulai mengembangkan Proyek PIR Perkebunan yang pertama pada tahun 1977 pada dua lokasi di Aceh dan Sumatera Selatan dengan budidaya karet. Pihak Bank Dunia memberikan sambutan yang sangat positip terhadap tekad Pemerintah, dan ini diwujudkan dalam bentuk bantuan pembiayaan bagi Proyek PIR Perkebunan yang pertama tersebut.'

Procedure for the NES Scheme

The structure of NES scheme is summarised in Figure 2-a. A company establishes a mill and its own nucleus estate (called *Inti* in Indonesian). Participating farmers organise

[31] Boeke defined 'social dualism' as 'the clashing of an imported social system with an indigenous social system of another style. Most frequently the imported social system is high capitalism' (Boeke 1953, p. 15).

cooperatives and contract with companies. Development banks, such as the World Bank, Asian Development Bank, or Indonesian Bank, provide credit to establish the participating farmers' smallholdings (*Plasma*) through channel banks to the nucleus estate companies (1). *Plasma* smallholdings are established by the companies around the nucleus estate (2). After harvest time arrives, the quality of the *Plasma* smallholdings is evaluated by the Directorate General of Estate Crops in the Ministry of Agriculture (3), and the National Land Office (*Badan Pertanahan National*) issues a certificate of private land ownership (*Hak Milik*) to the participating farmers[32](4). After the evaluation, the participants sign formal loan agreements with the channel bank and the name on the credit agreement is changed from the Directorate General of Estate Crops to each participating farmer. The *Plasma* smallholdings (typically two ha/household) are transferred from the company to the participants (5), and the channel bank holds the land certificates as collateral. The participants then begin managing the *Plasma* smallholdings. The participants sell their products to the contracted companies (6). The companies deduct 30% from the sales of the products as repayment (7). After the loans are fully repaid, the banks return the land certificates to the participants.

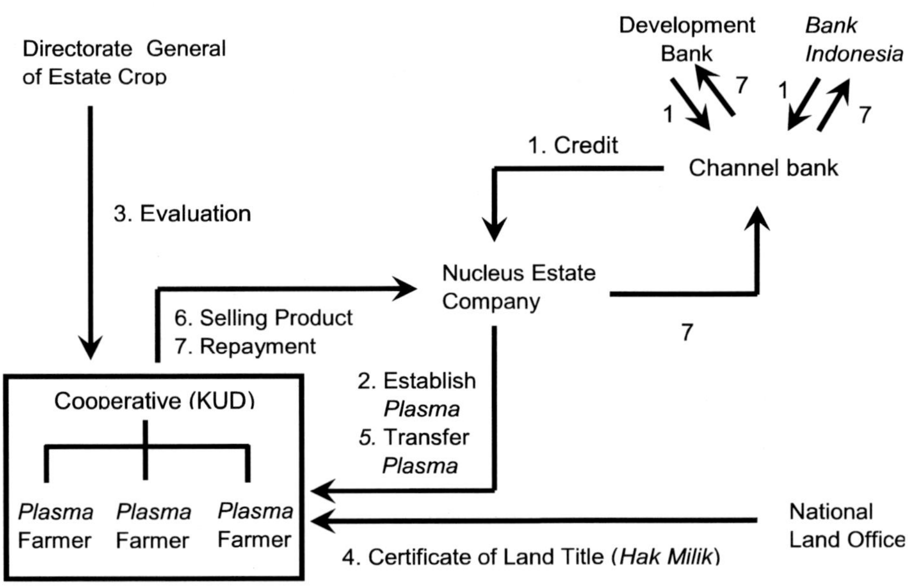

Figure 2-a. Structure of NES scheme.

Farmers who participate in NES schemes strongly depend on companies. Additionally, the local community must provide part of its traditional land to the company. A 25- to 35-year land use concession (*Hak Guna Usaha; HGU*) to the nucleus estate is issued by the National Land Office for the land, in which the local people agree to surrender their traditional land tenure. After the *HGU* expires, the land is returned to the nation[33], which means that the possibility that the land is returned to the people who have already surrendered land

[32] The cost for issuing a land certificate is covered by the NES project budget and is included in the debt of participating farmers when the *Plasma* smallholdings are transferred to participating farmers.

[33] This is noted in Articles 4 and 18 in the government decree of Republic of Indonesia No. 40 in 1996.

ownership is very low. This scheme leads to many social conflicts related to land disputes[34]. Companies sometimes bribe village heads or traditional law heads to obtain agreements for land expropriation.

Overview of NES Scheme Development

Table 1-a shows the development of NES programs up to 1999. Three NES programs are implemented by the central government: *PIR-BUN* (*PIR-Perkebunan*), *PIR-TRANS* (*yang dikaitkan dengan program Transmigrasi*) and *PIR-KKPA* (*Kredit Koperasi Primer untuk Angota*). The first generation program was *PIR-BUN*. The main perennial crops in this program were oil palms (43%) and rubber (48.5%). Coconut, tea and sugar corn occupied only 8.5% of the total area covered by this program. *PIR-BUN* was implemented primarily by state-owned companies (10 state-owned companies and 2 private companies). In 1986, *PIR-Trans*, the second generation of NES programs, began. *PIR-Trans* concentrated mainly on oil palms (91%). Oil palm expansion in *PIR-Trans* was conducted mainly by private companies (54 companies). Since 1995, *PIR-KKPA* has been implemented by 38 companies (Departemen Pertanian 1999).

Zen et al., (2006) reported that the performance of the initial *PIR* program is not good because of failed subsistence food crops, the government's decision to set FFB (Fresh Fruits Bunches) prices too low, and the 30% deduction for repayment. Many participants could not generate enough income and abandoned their lands, selling their plots to rich traders (ibid, p. 22). However, 'the settlers' situations usually improved greatly when their oil palms went into full production after 9-10 years, which was also a stage when they were normally able to complete both their principal and interest repayments. Although there were exceptions, most of the farmers in *Plasma* areas established up to the mid-1990s were doing well by the mid-2000s, although there was the usual variable performances characteristic of small farm agriculture' (ibid, p.23). On the other hand, many social and environmental problems occurred in the new era, such as large debts and frequent delinquency, and delays by mills in buying products (Marti 2008). Regarding the impact on the tropical rain forest, 70% of the oil palms established by 1999 were developed in forest areas (Anne 2000). Many companies became the targets of popular demonstrations after Suharto resigned in 1998, and the political power of the local people increased so that companies are now required to compensate them for land expropriation and improve their socioeconomic conditions (Zen et al., 2006 and Marti 2008). Government-sponsored NES programs declined after experiencing economic crises and political change from the end of the 1990s to the beginning of the 2000s.

Project Management Unit (PMU) *Scheme*

The PMU scheme project began in 1979. In this scheme, government project management units (PMU) are organised to support individual smallholders.

---

[34] As of January 2008, Sawit Watch was monitoring 513 active conflicts between companies and communities in Indonesia's oil palm plantation sector (Marti, 2008).

Table 1-a. Development of NES programs by 1999

Program	*PIR-BUN*					*PIR-TRANS*					*PIR-KKPA*				Total			
Plant year	1981-1993					1986-1998					1995-1999				-			
Financial resource [1]	WB, ADB, KfW, SFD, CDC, *Bank Indonesia*					*Bank Indonesia*					*Bank Indonesia*				-			
	Inucleus (ha)	*Plasma* (ha)	Total (ha)	Family	Class A+B	*Nucleus* (ha)	*Plasma* (ha)	Total (ha)	Family	Class A	Nucleus (ha)	*Plasma* (ha)	Family	Class	*Nucleus* (ha)	*Plasma* (ha)	Total (ha)	Family
Oil Palm	65,813	160,816	226,629	82,247	85%	151,670	394,334	546,005	167,847	91%	N/A	97,272	N/A	N/A	217,483	652,423	869,906	250,094
Rubber	83,323	172,268	255,591	74,804	58%	-	-	-	-	-	N/A	1033.22	N/A	N/A	83,323	173,301	256,624	74,804
Coconut	12458	13549	26,007	9795	53%	12,900	44,190	57,090	22,095	100%	-	-	-	N/A	25,358	57,739	83,097	31,890
Tea	829	9,631	10,461	24,156	52%	-	-	-	-	-	-	-	-	N/A	829	9,631	10,461	24,156
Sugar Corn	4,297	7,437	11,734	5,277	100%	-	-	-	-	-	-	-	-	N/A	4,297	7,437	11,734	5,277
Grand total	166,720	363,702	530,422	196,279	71%	164,570	438,524	603,095	189,942	92%	111,359	98,305	N/A	N/A	331,290	900,531	1,231,822	386,221

Note 1: WB: World Bank, ADB: Asian Development Bank, KfW: Kreditanstalt für Wiederaufbau, SFD: Saudi Funds of Development CDC: Commonwealth Development Corporation

Source: Rearranged Kawai (2011)

Table 1-b Development of PMU projects by 2002

Project [1]	PRPTE		SRDP I		SRDP II		SCDP		TCSDP-GOI		TCSDP		TCSSP		STCPP		ISDP		UFDP		EISCDP		S3TCDP		Total
Plant Year	1979-1984		1980-1986		1986-1990		N/A		1988-1990		1992-1998		1992-2000		1993-2002		N/A		1993-2002		N/A		N/A		-
Fund Resouce [2]	Bank Indonesia		WB		WB		WB		Bank Indonesia		WB		ADB		ADB		WB		ADB		IFAD		IFAD		-
Commodity	Area (ha)	Class Good [3]	Area (ha)	Class A+B	Area (ha)	Class A+B	Area (ha)	Class A+B	Area (ha)	Class A+B	Area (ha)	Class A+B	Area (ha)	Class A+B	Area (ha)	Class A+B	Area (ha)	Class A+B	Area (ha)	Class A+B	Area (ha)	Class A+B	Area (ha)	Class A+B	Area (ha)
Rubber	92,053	**57%**	46,500	**68%**	73,112	**70%**	-	**68%**	9,035	**58%**	71,025	**76%**	78,000	**N/A**	20,965	**N/A**	-		4,510	**N/A**	-		-		395,204
Coconut	147,461	**58%**	-	-	-	-	30,000	-	14,799	**45%**	41,032	**45%**	-		-		12,500	**N/A**	-		-		6,825	**N/A**	252,618
Coffee	62,731	**60%**	-	-	-	-	-	-	-	-	-	-	-		-		-		-		-		-		62,732
Tea	2,717	**34%**	-	-	-	-	-	-	-	-	-	-	19,925	**N/A**	-		-		-		-		-		22,642
Pepper	8,969	**60%**	-	-	-	-	-	-	-	-	-	-	-		-		-		-		-		-		8,970
Cacao	5,557	**42%**	-	-	-	-	-	-	-	-	-	-	-		-		-		-		-		-		5,557
Cashews	-	-	-	-	-	-	-	-	-	-	-	-	-		-		-		5,500	**N/A**	34,820	**N/A**	-		40,320
	319,488	**N.A**	46,500	**68%**	73,112	**70%**	30,000	**68%**	23,834	**50%**	112,057	**64%**	97,925	**N.A**	20,965	**N/A**	12,500	**N/A**	10,010	**N/A**	34,820	**N/A**	6,825	**N/A**	788,043

Note 1: PRPTE: Peremajaan, Rehabilitasi, dan Perluasan Tanaman Ekspor, SRDP: Smallholder Rubber Development Project, SCDP: Smallholder Coconut Development Project, TCSDP: Tree Crops Smallholder Development Project, TCSSP: Tree Crop Smallholder Sector Project, STCPP: Smallholder Tree Crop Processing Project, ISDP: Integrated Swamps Development Project, UFDP: Upland Farmers Development Project, EISCDP: Eastern Islands Smallholder Cashew Development Project, S3TCDP: South Sumatera Smallholder Tree Crops Development Project

Note 2: WB: World Bank, ADB: Asian Development Bank, IFAD: International Fund for Agricultural Development

Note 3: The category "Good" in PRPTE is a temporary evaluation result. It did not follow the evaluation standard of the Department of Agriculture (DP, 1986).

Source: Rearranged Kawai (2011)

Characteristics of PMU Scheme

The main characteristics of the PMU scheme are as follows:

1) Implemented on existing traditional farms, secondary forests, and fallow and degraded forests around traditional farms.
2) Training smallholders as modern and independent smallholding managers.
3) Establishing new economic spheres around PMUs. One PMU manages 3,000-5,000 ha of smallholdings[1].
4) Supporting both transmigrants and local people.

Concept of PMU Scheme

A government report explained the concept of the PMU scheme as follows.

'The PMU scheme is intended to support farmers in very disparate locations. The farmers participating in this project are supported by the PMUs with advanced technology in establishing their plantation. It is then oriented to allow these farmers to organize themselves by creating famer groups. With this action, it is expected that farmers can establish their plantation, manage their farming continually, and enjoy the profits so that they become independent farmers.

Thus, it can be said simply that the concept of the PMU is substantial enough because:

- It allows existing famers to play a role in the acquisition of foreign currency from non-oil and natural gas sectors.
- It significantly educates farmers regarding the development of modern farming, which satisfies technical bank requirements and realises further increases in productivity.
- Every year, the total number of farmers covered by the PMU increase

With this scheme, in fact, it is important to include the principle of equalization, which itself will grow in accordance with the increase of productivity.

In line with productivity increases and income growth and development, the PMU areas become the centre of the economy (Departemen Pertanian 1984, p. 4: translated by the authors).

Pola UPP dimaksudkan untuk menjangkau pekebun yang sudah ada dengan lokasi yang sangat tersebar. Petani peserta proyek dibantu oleh Unit Pelaksanaan Proyek dalam membangun kebunnya dengan teknologi maju kemudian diarahkan agar petani tersebut mengorganisir diri dalam bentuk kelembagaan petani. Dengan upaya ini diharapkan petani dapat membangun kebunnya dan kemudian mengelola usaha taninya secara kontinyu dan menguntungkan sehingga menjadi petani yang mandiri.

Dengan demikian secara singkat dapat dikatakan bahwa konsepsi pola UPP cukup menndasar, karena:

- *Diarahkan kepada pekebun yang sudah ada yang selama ini telah berperan di dalam pendapatan devisa non migas.*
- *Secara massal mendidik petani ke arah bentuk usaha tani yang berciri modern, memenuhi persyaratan bank teknis, dan adanya produktivitas yang semakin meningkat.*
- *Dari tahun ke tahun jumlah petani yang dijangkau semakin meningkat.*

[1] The economically feasible size of the PMU for rubber is 3,000-5,000 ha per PMU and 2,000-3,000 ha per PMU for coconut (Ministry of Forestry and Estate Crops, 2000).

Dengan pola ini pada hakekatnya ingin dijangkau azas pemerataan yang dengan sendirinya akan diikuti adanya pertumbuhan sejalan dengan meningkatnya produktivitas.

Sejalan dengan meningkatnya produktivitas, meningkatnya pendapatan dan perkembangannya wilayah UPP menjadi sentra-sentra ekonomi

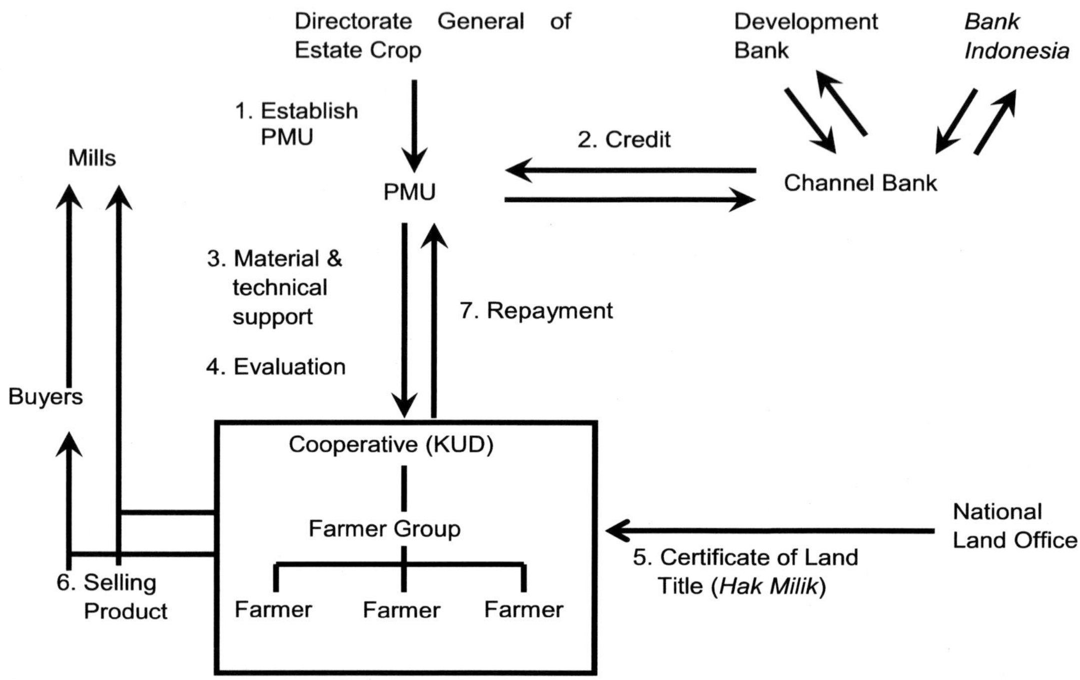

Figure 2-b. Structure of PMU scheme.

Structure of PMU Scheme

The structure of the PMU schemes is summarised in Figure 2-b. The Directorate General of Estate Crops in the Ministry of Agriculture establishes a PMU in each project area (1). Participants organise farmer groups and/or cooperatives. They prepare their land for participation in the PMU project. The development banks or *Bank Indonesia* provides credit to the PMU (Directorate General of Estate Crops) through a channel bank (2). PMU use the money to buy materials, such as high-yield clone seedlings, fertiliser, agrichemicals, and farming tools. The participants establish smallholdings with material and technical support from the PMU (3). When the crops mature (at harvest time), each smallholding is evaluated by a team from the Directorate General of Estate Crops (4). The amount of credit is decided based on the result of the evaluation. If the crops fail to grow, the government covers the loss (repaying the credit) so that the participants are protected from risk. The National Land Office issues land certificates (*Hak Milik*)[2] to the smallholdings (5). After evaluation, the participants sign formal loan agreements with the channel bank and the name on the credit document is changed from the Directorate General of Estate Crops to the name of the participating farmer. The channel bank holds the land certificates as collateral. The participants sell the products to

[2] The cost of issuing the certificate of land title is covered by the project budget and it is included in the debt of participating farmers.

buyers or mills directly (6). The PMU manages the participants' loan repayment (collecting it in instalments) (7). After finishing repayment, the banks return the land certificates to the participants. The participants become the owners and managers of independent high-yield modern smallholdings.

The main differences between the PMU and NES schemes are 1) the participants establish their smallholdings by themselves with support from the PMU, 2) there is no land expropriation, and 3) the participants can sell their products to buyers and mills freely (but access to the market is not guaranteed). Most importantly, the PMU scheme participants become independent smallholding owners and managers, while NES scheme participants strongly depend on the companies.

Overview of PMU Scheme Project

The authors identified 12 PMU projects, shown in Table 1-b. The first PMU project was *PRPTE* (*Peremajaan Rehabilitasi dan Perluasan Tanaman Ekspor*), funded by the national budget. After *PRPTE*, most PMU projects were funded by development banks, such as the World Bank, Asian Development Bank and International Fund for Agricultural Development from the 1980s to the 1990s[3]. The PMU scheme, however, was not used for oil palms because oil palms must be developed near mills[4]. PMU project sites are sometimes far from markets or mills; consequently, they may be suitable for crops whose products can remain stable for long periods of time, such as rubber.

Barlow and Tomichi (1991) analysed the performance of *PRPTE* and concluded that 'the outcome of the *PRPTE* scheme was very disappointing and this could be attributed largely to shortfalls in both budgeted material input supplies and technical guidance' (ibid, p. 46). They reported that only 26% of rubber plantations in *PRPTE* was classified as 'satisfactory' regarding its ability to produce reasonable yields. However, the report of the Ministry of agriculture classified 40% to 60% of the rubber as 'good' (Table 1-b). The report, however, recognised that the evaluation only partly satisfied the standard. The 'satisfactory rate' of PMU projects increased by 68% in the Smallholders' Rubber Development Project (SRDP) and Smallholders' Coconut Development Project (SCDP) — which is supported by the World Bank — due to greater material input and more scrupulous management (ibid, p. 47). The rate also increased by 76% at the Tree Crops Smallholder Development Project (TCSDP) conducted in the 1990s (Ministry of Forestry and Estate Crops 2000). The performance of the PMU schemes seems to be increasingly improving. The report of the TCSDP and Tree Crop Smallholder Sector Project (TCSSP) discusses the socioeconomic impact of PMU projects, such as creating stable employment opportunities, including for women, and improvements in peoples' lives, including better housing and the opportunity to send their children to school; the projects have also given the people a sense of accomplishment (Ministry of Forestry and

---

[3] PMU project was implemented in Aceh, North Sumatra, West Sumatra, Riau, Jambi, South Sumatra, Bengkulu, Lampung, West Jawa, Central Jawa, Yogyakarta, East Jawa, West Kalimantan, Central Kalimantan, South Kalimantan, East Kalimantan, North Sulawesi, Central Sulawesi, South Sulawesi, South East Sulawesi, Bali, West Nusa Tenggara, East Nusa Tenggara, Maluku and Irian Jaya (Departemen Pertanian 1984; Departemen Pertanian 1985; Barlow and Tomichi 1991; Ministry of Forestry and Estate Crops, 2000; Asian Development Bank 2002; Asian Development Bank 2004 a; Asian Development Bank 2004 b; Asian Development Bank 2005).

[4] Palm oil fresh fruit bunches must be milled within 24 hours of harvest to avoid deterioration in quality.

Estate Crops 2000: Asian Development Bank 2002). In contrast to the NES scheme, little negative social impact has been reported. Local people use their own land and do not need to surrender their customary land. Regarding the environmental impact, although one PMU manages a relatively large 3,000 ha - 5,000 ha smallholding, each operational block is relatively small (100 - 300 ha in *PRPTE* and 200 – 300 ha in SRDP), and the blocks are scattered (Barlow and Jayasuriya 1984; Barlow and Tomochi 1991). The environmental impact, then, is less than that of the NES schemes. Additionally, the largest area to which the PMU scheme was applied was on existing farmland, bush and *alang-alang* (*Imperata cylindrical*), rather than natural forest (Ministry of Forestry and Estate Crops 2000; Asian Development Bank 2002).

The main challenge of PMU schemes is the low repayment rate (Barlow and Tomichi 1991; Asian Development Bank 2002). According to a Ministry of Agriculture report, the average repayment rate of eight PMU projects was 35.8% as of 2003 (Departemen Pertanian 2006). There is, however, a flexible PMU type scheme that has addressed the low repayment problem at a certain level in the Paser district in East Kalimantan. In this case, the PMU scheme was applied to oil palm smallholdings. The PMU contracts with oil palm companies to sell products from smallholders. The financial resources come from the local government budget rather than a development bank. The credit size is limited to high-yield seedling expenses and the administration fee for the land certificate. Fertilisers and agricultural chemicals are provided gratis by the local government but only for the first and second years. After this period, the participants must buy the fertiliser and agrichemicals themselves. The participants repay the loan to the PMU, and the local government provides credit to the subsequent participants. Although this local government's smallholder development scheme is not official PMU scheme program implemented by central government, this chapter regards this scheme as "PMU type" scheme in the broad sense because this scheme has some characters of PMU scheme such as (1) project management unit give participants technical and material supports, and participants establish smallholdings by themselves, (2) the scheme does not include direct companies' support for participants[5], (3) the scheme provides credit to participants to establish smallholdings and (4) the private land ownership is provided to participants. The details of this flexible PMU type scheme are analysed in the following section.

Partial Support

In addition to the NES and PMU schemes, Indonesia has given partial support to smallholders. This partial support is a complementary scheme and has been applied in areas where neither the NES nor PMU schemes have been used. Partial support provides some support, such as 'only' high-yield seedlings, fertiliser, agrichemicals or farm tools. This support creates an incentive for farmers to increase their farms' production. Farmer groups are organised, and government agricultural instructors provide technical support. Credit and land certificate preparation is usually not included in the partial support. Most activities are

[5] There are some commercial partnership between PMU and companies such as that PMU purchase high-yield seedlings from companies and sell smallholders' products to them. In this case, PMU play a role as mediator agency between companies and smallholders.

conducted through the farmers' own efforts. The shortcomings of partial support are budget shortages and limitations on technical support and information distribution.

Non-Support Smallholdings

In this paper, smallholdings that are established by farmers without any support from companies or the government are called non-support smallholdings.

The Impact of NES and PMU Schemes in East Kalimantan

This section seeks to clarify the impact of the NES and PMU schemes in East Kalimantan. The authors used the data of three sites — that is, the Middle-Upper Mahakam region in the Mahakam Ulu and West Kutai districts, the Sendawar region in the West Kutai District, and the Paser district.

The Economic Conditions of the Middle-Upper Mahakam Region

As Table 2 shows, 33.4% of the sample households lived under the Indonesian poverty line of US \$0.60. Many households lost their cash income as a result of the decline of the commercial logging industry and the prohibition of illegal logging by the government.

The average cash income per year from 'Forestry' was 50,000 Rupiah; 'Company' was 0 and 'Total' was only 5.4 million rupiah. People earned little from 'fish' and 'labour'[6]. Some households could not meet educational and medical expenses. In this region, 20 million rupiah per year per household was established as the minimal cash income to live without significant deficiencies. The cash income resources to meet this level were 'government,' such as working as public servants, and 'merchant,' such as shopkeepers. Opportunities to obtain these jobs, however, were limited for many households in this region.

Table 2. Economic condition of sample households in each area and village (%)

Poverty Line ($US)	N	Very Low < 0.60	Low 0.60 - 1.08	Middle 1.08 - 2.00	High 2.00 <	Total
Middle-Upper Mahakam	48	33.4	20.8	33.3	12.5	100.0
T Village	25	0.0	24.0	40.0	36.0	100.0
G Village	20	0.0	5.0	10.0	85.0	100.0
N Village	20	10.0	15.0	25.0	50.0	100.0

Source: Based on Kawai and Inoue (2010)

6 "Labour" is wages from working on a neighbour's farm, such as swidden. This is a type of labour exchange in the village and has a redistributive function of cash income for the poor.

Table 3-a. Average cash income of sample households per year in Middle-Upper Mahakam region (10,000 Rp)

	N	Rice	Vegetable	Fruit	Fish	Hunting	Livestock	Forestry	Rubber	Labour	Government	Company	Craftsman	Merchant	Others	Total
Very Low	16	5	12	7	163	1	12	5	0	185	45	0	63	0	47	544
Low	10	0	4	41	152	55	6	36	213	102	88	0	110	169	386	1,362
Middle	16	11	39	54	117	38	10	35	176	136	648	153	38	424	160	2,038
High	6	0	0	326	156	533	103	250	406	0	1,481	450	0	764	381	4,849
Total	48	6	18	70	144	91	21	52	154	128	434	107	57	272	197	1,750

Source: Based on Kawai and Inoue (2010)

Table 3-b. Average cash income of sample households per year in T village (10,000 Rp)

	N	Rice	Vegetable	Fruit	Fish	Hunting	Livestock	Forestry	Oil Palm	Labour	Government	Company	Craftsman	Merchant	Others	Total
Very Low	0	0	0	0	0	0	0	0	0	0	0	0	0	0	0	0
Low	6	66	117	3	0	0	0	0	222	0	124	666	63	0	0	1,260
Middle	10	45	9	5	10	0	19	0	1,133	0	27	385	174	183	12	2,001
High	9	0	0	67	0	0	85	0	2,214	0	1,182	870	360	1,568	196	6,541
Total	25	34	32	27	4	0	38	0	1,303	0	466	627	200	637	75	3,443

Source: Based on Kawai and Inoue (2010)

Table 3—c. Average cash income of sample households per year in G village (10,000 Rp)

	N	Rice	Vegetable	Fruit	Fish	Hunting	Livestock	Forestry	Oil Palm	Labour	Government	Company	Craftsman	Merchant	Others	Total
Very Low	0	0	0	0	0	0	0	0	0	0	0	0	0	0	0	0
Low	1	0	0	0	0	0	0	0	1,128	0	0	0	0	0	0	1,128
Middle	2	0	0	0	0	0	0	0	1,822	0	0	0	0	0	0	1,822
High	17	15	1	14	71	0	18	0	3,653	0	222	427	261	676	71	5,429
Total	20	13	1	12	60	0	15	0	3,344	0	189	363	222	514	121	4,853

Source: Based on Kawai and Inoue (2010)

Table 3—d. Average cash income of sample households per year in N village (10,000 Rp)

	N	Rice	Vegetable	Fruit	Fish	Hunting	Livestock	Forestry	Rubber	Labour	Government	Company	Craftsman	Merchant	Others	Total
Very Low	2	0	0	0	0	0	0	0	731	0	0	0	0	0	0	731
Low	3	0	0	0	0	0	0	0	1,732	0	0	0	0	0	0	1,732
Middle	5	0	0	0	0	0	31	0	2,356	0	0	156	0	0	0	2,543
High	10	0	0	0	0	0	116	0	3,156	10	224	360	0	0	0	3,866
Total	20	0	0	0	0	0	66	0	2,498	5	112	219	0	0	0	2,900

Source: Based on Kawai and Inoue (2010)

Note 1: "Rice" is selling rice. "Vegetable" is selling vegetables, such as cassava, eggplants, cucumbers, peanuts, string beans, corn, green onions, sweet potatos. "Fruit" is bananas, coconuts, durians, papayas, pineapples, mangoes, rambutans, and langsat. "Fishery" is selling fish. "Hunting" is selling the meat of hunted boars and deer. "Livestock" is selling the meat of boars and chickens. "Forestry" is cash income from participating in logging activities and selling timber and resin (Damar). "Rubber" is selling latex. "Oil Palm" is selling fresh fruit bunches of oil palms. "Labour" is wages from work in neighbours' swidden. "Government" is working as a public servant or temporary worker, such asa staff member of a sub-district or village, elementary school or junior high school teacher. "Company" is wages from working for a company. "Craftsman" is a carpenter or someone who makes and sells handicrafts, such as rattan work, beadwork, or needlework. "Merchant" is a small business worker, such as a shopkeeper, seller of basic goods, telephone station owner, fuel seller, middleman or carrier of FFB in T village and G village. "Others" is interest income from CU; wages from cooperatives; sales of swallows' nests, gold dust, or land; and family support.

Table 4. Average self-sufficiency rate of sample households per year in each region and village (%) [1)]

Region and village	N [2)]	Rice	Vegetable	Fish	Meat	Total
Middle-Upper Mahakam	47	57	61	58	65	59
T Village	21	54	27	28	15	33
G Village	20	6	1	11	18	8
N Village	19	49	24	10	31	33

Note 1: Self-sufficiency rate is calculated by (estimated price of the product consumed per year) / (total cash expenditure for the product per year + estimated price of the product consumed per year) * 100.

Note 2: Data on one household in the Middle-Upper Mahakam region; four households in T village and one household in N village were removed from the analysis because the data on their expenditures were insufficient.

Source: Based on Kawai and Inoue (2010)

However, traditional swidden agriculture, fishing, hunting and other non-timber forest product (NTFP) -gathering activities still continued in the Middle-Upper Mahakam region. Table 4 shows the self-sufficiency rate of each product. Although the self-sufficiency rate of rice was 57%, it is 100% in a typical year. Rice production decreased drastically in 2007 because of a long drought. Many families had to buy rice from external sources. The self-sufficiency rates of vegetables, fish and meat were 61%, 58% and 65%, respectively. Additionally, local people could use timber for their own needs, such as houses and boats, but selling timber to outside markets, such as Melak town (one of the large towns in the West Kutai district) and Samarinda city (the provincial capital), was restricted by the government[1]. Firewood was also acquired from customary forests. Forest-based traditional livelihoods play an important role in the subsistence of the local people.

At the time of this research, although people maintained forest-based traditional livelihoods, alternative cash income resources were generally necessary to offset the declining commercial logging industry revenue in the Middle-Upper Mahakam region.

Oil Palm Development in the Paser District

As of 2008, the total area of oil palm plantations in the Paser district was 95,822 ha. The nucleus estates established by state-owned companies covered 13,440 ha, the nucleus estates established by private companies covered 46,475 ha, the *Plasma* smallholdings established by state-owned companies (NES scheme) covered 24,854 ha, the smallholdings supported by the local government (PMU type and partial support scheme) covered 9,345 ha and non-support oil palm smallholdings covered 1,708 ha. The data on non-support oil palm smallholdings, however, were not correct. The estate office staff stated that there are many non-support oil palm smallholdings that are not counted by or registered with the estate office.

In T and G villages, the majority population is the indigenous Paser people. As of 2008, the population of T village was 3,156 individuals and 811 families. Seventy-five percent were

[1] The main illegal logging government checkpoints exist in the downstream area, such as in Long Iram and Melak town; as a result, timber utilisation by local people within the villages of the Middle-Upper Mahakam region is tacitly approved.

Paser people. The oil palm plantations have been developed since 1991 by a state-owned company. According to the village statistics, the area of the village land was 9,700 ha[2]. The expropriated land designated as the nucleus estates of state-owned companies measured 1,500 ha, so a relatively large area of land remained for the villagers. The estimated size of villagers' oil palm smallholdings were approximately 750 ha based on village statistical data. There were also paddy fields and wetlands in the lowland area.

In G village, the population was 3,976 people and 1,071 families in 2009. According to village statistics in 2006, 42% are Paser people, 37% are Javanese and 21% belong to other ethnic groups. In 1983, a state-owned company estate was developed. The area of G village was 10,500 ha[3]. However, 5,300 ha was state-owned forest land. Nucleus estate established by state-owned companies measured 2,800 ha, and nucleus estates established by private company measured 940 ha, which means that many parts of village areas were in state-owned forest or expropriated by companies. Only 1,460 ha of land remained for the villagers. Villagers' smallholdings were 809 ha. Although there were a few hectares of coconut, coffee, pepper and rubber smallholdings, most of them were oil palm smallholdings. Paddy fields measured 12 ha. Swidden/rain-fed farm land was 202 ha. There were 160 ha of wetland.

In both villages, indigenous people were initially discriminated against in oil palm plantation development. At that time, which was during the Suharto regime, state-owned companies expropriated local peoples' customary land with the support of the military. Villagers cultivated rice in both swidden and paddy fields. The former village leader of T village stated, 'at that time, villagers obtained cash income from selling bananas, rattan and latex. However, the amount was small.' After 1991, the Paser district government began partial-support schemes (*P4: Proyek Peningkatan Produksi Perkebunan*[4]) to establish oil palm smallholdings, and after 1993, the provincial government began to support the establishment of oil palm smallholdings for the local people through the PMU type scheme (*PIR-Swadaya*). The participants, however, were limited. The *Plasma* smallholdings for the local people were also not developed despite promises made by the state-owned company. In 2000, after democratisation, some villages, including T and G villages, demonstrated against the state-owned company. They then received *Plasma* smallholdings. After that, the state-owned company established its own NES scheme program called *PIR-KKPA*[5]. Many local people joined the program and received *Plasma* smallholdings. The state-owned company also hired many local people[6].

The results of the household survey shows that the average area — based on 17 households[7] who have oil palm smallholdings in T village — was 2.3 ha and that the

[2] The figure changes depending on the data resources (the area is 7,910 ha according to the data from the statistics office (*Badan Pusat Statistik*) of the Paser district).

[3] The area is based on the statistical data of the village. The figure changes depending on the data resources (the area is 9,051 ha according to the statistics office of Paser district).

[4] This is a partial-support program of the local government of *Paser* District. The local government provided high-yield oil palm seedlings, fertiliser and agrichemicals.

[5] Although the name *PIR-KKPA* is the same as that of the *NES* program implemented by the central government (Table 1-a) in the 1990s, it is a different program. The *PIR-KKPA* program is *PTPN XIII*'s own *NES* program.

[6] According to the state-owned company staff, local workers comprise approximately 50% of total workers in both the nucleus estate and the factories.

[7] Twelve households have Plasma (PIR-KKPA) oil palm smallholdings supported by the NES scheme. Six households have partial support (*P4: Proyek Peningkatan Produksi Perkebunan*) oil palm smallholdings supported by the local government. Three households have non-support oil palm smallholdings, and seven households do not have oil palm smallholdings. Three households have plural scheme oil palm smallholdings.

households earned 18.4 million rupiah in one year. The average area of 20 households[8] with oil palm smallholdings in G village was 3.3 ha, and they earned 32.8 million rupiah in one year. The advantage of oil palm cultivation was that it can be developed as a side business; there are only two harvest periods each month; and fertilisation and weeding are needed only 2-3 times per year. Some smallholders hire workers for the harvest. Smallholders then need not devote much time to oil palm maintenance. Thirty-one sample households had a side business. The sources of high cash income in Tables 3-b and 3-c were 'Company,' e.g., working for the oil palm company; 'Government,' such as working as a public servant or village government staff member; and 'Merchant,' such as working as an FFB carrier, middleman, or shopkeeper. Only six sample households earned cash income only from oil palm smallholdings. The 'Low'-income households with oil palms comprised only three households in both villages. Therefore, the household survey data show that oil palm smallholdings are quite effective in terms of alleviating poverty. However, four in eight households who do not have oil palm smallholdings remained under the poverty line (US $1.08). These households did not have land for oil palm smallholdings. An NES scheme by the state-owned company *PIR-KKPA* required participants to have their own land, so the villagers without land could not participate. This feature is different from *PIR-BUN* and *PIR-TRANS*, the earlier central NES scheme projects.

Table 5 compares the production, revenue and expenditures of each crop and scheme per hectare. The highest production is acquired in nucleus estate. Next, *Plasma*, PMU type, and partial support smallholdings show almost the same relatively high productivities. These schemes used high-yield seedlings supplied by state-owned companies. However, the production of non-support smallholdings was low. Their seedling origin was not clear. The important point in this table is that the profits of PMU type and partial support smallholdings exceed those of *Plasma* smallholdings due to transportation costs and credit amounts. The total credit amount of *Plasma* smallholdings was approximately 30 million rupiah per hectare. In contrast, the credit amount in PMU type schemes was only five million rupiah per hectare. In partial-support schemes, there was no credit. In PMU type and partial support schemes, participants work the land themselves and do not use labourers to save on costs. Additionally, these smallholdings were established around the villages and the main road to save on transportation costs. However, *Plasma* smallholdings were situated in areas farther from residences, and the transportation costs increased. The advantage of PMU type and partial-support schemes and is that the participants can choose a market with both middlemen and mills. In *Plasma* smallholdings, participants may only sell to a contracted company. Four in seven households with both PMU type scheme and *Plasma* smallholdings stated that the PMU scheme is better than the NES scheme because the credit amount is small for the NES scheme. The remaining three households stated that both have advantages. The advantages of the *Plasma* smallholdings are 'High production' and 'Company builds a road from the main roads.' Eighteen participants in 20 households with PMU type scheme smallholdings had finished repaying their loans by time of this study. The problem was in delays in issuing the land certificates. Only seven households received land certificates, two households received land certificates for only part of their smallholdings and nine households had not yet received any land certificate.

[8] All sample households had government flexible PMU type scheme (*PIR-SWADAYA*) oil palm smallholdings. Seven households have *Plasma* oil palm smallholdings, and four households have non-supported oil palm smallholdings in addition to PMU scheme oil palm smallholdings.

Table 5. Comparison of production, revenue and expenditure of each perennial crop and scheme per ha as of 2008 [1), 2)]

Perennial Crops	Oil Palm					Rubber	
Scheme [3), 4)]	Nucleus Estate	Plasma (PIR-KKPA)	PMU (PIR-SWADAYA)	Partial Support (P4)	Non-Support	PMU (PRPTE/TCSSP)	Non-Support
N (Household)	-	19	20	6	6	18	9
Areal (ha/Household)	-	2.2	2.3	0.8	1.9	1.4	1.9
Production (ton/ha/year)	16.7	13.3	12.0	11.2	7.3	2.6	1.2
Revenue (10,000 Rp/ha/Year) [5)]	**2,166**	**1,731**	**1,558**	**1,448**	**951**	**1,779**	**810**
Fertilizer	N/A	99	91	168	126	28	1
Pesticide	N/A	32	42	9	36	1	5
Labour [6)]	N/A	114	114	47	115	41	3
Transportation [7)]	N/A	201	52	28	34	0	0
Loss from selling to middleman [8)]	N/A	17	77	18	18	0	0
Repayment	N/A	519	0 [9)]	0	0	0 [10)]	0
Total Expenditure (10,000 Rp/ha/year)	**N/A**	**982**	**376**	**270**	**330**	**70**	**9**
Profit (10,000 Rp/ha)	**N/A**	**749 (1,268)** [11)]	**1,206**	**1,178**	**621**	**1,708**	**801**

Note 1: For oil palms, the data of 17 households in T village and 20 households in G village are used. For the PMU scheme, rubber smallholdings and data from 18 households in N village are used. For non-support rubber smallholdings, the data of 9 households in 6 villages around the Jempang Lake region (Jempang Siruq Ngurai and the Bongan sub-district) in West Kutai district are used.

Note 2: The planting year of each smallholding is different, which might affect the variability of production.

Note 3: One hundred twenty-eight oil palms per ha in Paser district and 525 rubber trees per ha in West Kutai district were planted in PMU scheme smallholdings. The number of plants per ha in non-support oil palm and rubber smallholdings is unclear.

Note 4: The name in parentheses is the project name in the field.

Note 5: Revenue is calculated by the (average price per year per kg)*(production per year per ha). The average FBB price per year is 1,297 Rp/kg, which was calculated by the local government in the Paser district. The average latex sale price per year for a middleman in N village is 6,783 Rp/kg.

Note 6: Most labourers are employed by the sample households in oil palm smallholdings for harvest work. The price is 150,000 Rp per ton per capita at the time of this research. In the case of rubber smallholdings, most labourers are hired for weeding. The price is 50,000 Rp per day per capita.

Note 7: This is the transportation cost for products from the smallholdings to the mill. The cost depends on the distance and road conditions. In T village, the cost is 50,000 Rp per ton in the area within 10 km from the mill, and 200,000-300,000 Rp per ton in the area that is 30-40 km from the mill. If the road becomes muddy in the rainy season, the cost increases.

Note 8: This is the calculated loss in case smallholders sell their products to middlemen rather than through the official route. Typically, middlemen buy products from smallholders at prices 200-300 Rp per kg lower than companies. However, smallholders still sell products to middlemen to obtain cash immediately because state-owned companies pay only once a month. Some Plasma owners sell their products to middlemen to avoid repayment. In the case of rubber in the West Kutai district, the only route to sell latex is through middlemen; consequently, this cost is not calculated.

Note 9: At the time of this research, 18 in 20 households had finished repayment, and the other 2 households had begun repayment.

Note 10: Two households participating in PRPTE have already finished repayment. Repayment of TCSSP was delayed because the PMU merged with the district estate office after decentralisation. The repayment began in September 2008. The total amount of credit is approximately 7 million Rp per ha, including both the principal and interest. The repayment amount is lower than that of the Estate Revitalisation Program because the Rupiah collapsed after the Asian economic crisis in 1997. At the time of this research, there are still no sample households that have started repayment.

Note 11: The parenthetical number is the profit after repayment is finished.

Source: Based on Kawai and Inoue (2011)

From the viewpoint of the companies, managing *Plasma* smallholdings is difficult. Some staff of state-owned companies explained that although the companies provide technical support regarding optimal fertilisation and weeding, they cannot force participants to follow these instructions. The production of the sample 19 households' *Plasma* smallholdings ranges from 8 to 20.3 tons per hectare per year. The fertilisation costs are 0 to 3.5 million rupiah per year. The weeding costs are 0 to 1.4 million rupiah per year. There is significant variation among the smallholdings. A participant of the *Plasma* smallholdings scheme stated that 'an advantage of oil palms is that they bear FFBs without much maintenance. Low production (compared with production of the nucleus estate) is no problem because we have a sub-business.' Clearly, interests differ between the companies and smallholders. Each smallholder invests in his or her *Plasma* smallholding based on his or her own needs. Smallholders need not maximise profit just because the company wants to do so.

Regarding the land expropriation, 15%-19% of the land in T village and 36% of the land in G village was expropriated. Because the oil palm program started later and many hectares of wetland existed in T village, the land expropriation is limited, and paddy fields and oil palm plantations can coexist. Twelve in 17 sample households with oil palm smallholdings also cultivated paddy fields, and eight households cultivated vegetables. Notably, the self-sufficiency rate of rice in T villages is 54% (Table 4). However, in G village, oil palm monoculture proceeded. Few rice paddy fields and vegetable farms existed. Only two in 20 sample households in G village cultivated rice paddies and vegetables. The land for new cultivation was limited. According to village statistical data, only 160 ha of wetland remained uncultivated, aside from residences and state-owned forest area (*Kawasan Hutan*). This issue had become critical for the preparation of the next generation's land. It is important to consider how much traditional land is provided to companies if they grow oil palms.

Regarding the environmental impact, some villagers stated that after the oil palm plantations were developed, the groundwater often dried up. However, the impact on forest degradation in this area was not critical because villagers stated that the forest in this region was already degraded before the oil palm plantation existed.

Rubber Plantation Development in the Sendawar Region in West Kutai District

With 1,208 ha in PRPTE (1981-1983) and 9,179 ha in TCSSP (1992-2001), in total, 10,387 ha of high-yielding rubber smallholdings were developed by the PMU scheme in the Sendawar region.

According to the 2009 sub-district statistical data, the population of N village is 975, and the number of households totalled 244. The area of the village was 3,277 ha. The majority group is the Tonyoi Dayak people. In N village, 34 ha from *PRPTE*, 474 ha from TCSSP, and a total of 508 ha of high-yield rubber smallholdings were established, occupying 15.5% of the village area, Table 3-d shows the average cash income of the N village sample households. In 2008, the cash income from rubber smallholdings comprised 86% of the total cash income. Two 'very low'-income households did not join the PMU scheme. Fifteen in 18 sample households that joined the PMU type project were in the middle or rich category. Before the PMU type project started, the villagers' main occupations were slash-and-burn farming, cultivating vegetables, fishing, and hunting in the forest. The sources of cash income included unstable day labour and selling rattan. Villagers stated that the quality of life could not increase when they were only engaged in traditional farming. Then, after the PMU type scheme rubber project began, participants had stable cash incomes and could fulfil most of their needs, such as food, education, motorcycles, electrical appliances, furniture, and houses. According to villagers, the advantage of rubber smallholdings is that participants can earn stable cash income with only two hours of work every morning. The participants can work in other jobs, such as civil service and day labour, and they also have time to rest in the daytime. Villagers regard the lifestyle change through the PMU type project positively. In 2008, the profits from the PMU type rubber smallholdings per hectare exceeded those from the PMU type oil palm smallholdings (Table 5).

The challenges of the PMU type rubber scheme in this village were the further growth of rubber monoculture and the shortage of new land for cultivation. In N village, villagers who realised the profitability of high-yield rubber smallholdings developed competing new rubber smallholdings. Then, a land shortage occurred in this high-population density region. The reason that the self-sufficiency rate of rice was relatively high (49%) is that the households with land cultivate paddy fields that will later be converted into new rubber smallholdings. The other challenge was the repayment rate of the TCSSP project. The success rate of *PRPTE* project in this region was good (89%) (Departemen Pertanian 1985). Most participants finished repayment quickly to obtain their land certificates[1]. However, the repayment of TCSSP was delayed because PMU was integrated into the estate office of the West Kutai district after decentralisation. The repayment began in September 2008. Another challenge was low seedling quality in TCSSP. According to some agricultural office staff members in this area, the seedlings for *PRPTE* were high quality and these seedlings were obtained from Java by the PMU type staff directly, while the seedlings for TCSSP were prepared by contract companies. They said it was difficult to check the quality of seedlings prepared by contract companies because such seedlings had already been grafted in the places of production such as Medan, Palembang and Banjarmasin. Monopolisation of the local latex-selling market by buyers from Banjarmasin was also a challenge.

Alternative Development Strategies for Oil Palm Plantations

What type of development strategies can be considered based on the above results in pre-development regions such as the Middle-Upper Mahakam region? This paper considers two alternatives: 'industrialisation based on oil palm plantations with an NES scheme' and 'moderate industrialisation based on rubber smallholdings with a PMU scheme.'

Industrialisation Based on Oil Palm Plantations with the NES Scheme

Inoue (1994) coined the term 'incomplete industrialisation' for the phenomenon of people from different regions adapting industrialisation, such as the Javanese, who benefited from industrialisation (e.g., *HTI*, industrial afforestation), whereas the local people who live in forest areas experience few advantages or disadvantages from industrialisation. At first, oil palm development in the Paser district proceeded as incomplete industrialisation and excluded the local people. After democratisation, however, local people began participating in oil palm development and also started obtaining benefits. Oil palms could encourage job

[1] A senior officer of the agricultural office of this region suggested the dedicated support of the agricultural instructors to farmers at that time and the incentive to get land certificates were the key of the success.

creation, such as wage labourers in companies and transporters or middlemen of FFBs. The economic ripple effects of oil palms are very attractive to the local people.

There are, however, some risks. In the estate revitalisation program (*Program Revitalisasi Perkebunan*) [2]' launched by the central government in 2006, a new NES scheme called *PIR-PSM* (*Pola Satu Namajemen*) has been implemented. Companies manage *Plasma* smallholdings from planting time to replanting time (approximately 25 years) entirely to redress the differences between the nucleus estate and *Plasma* smallholdings and try to maximise productivity and efficiency. *PIR-PSM* removes participant smallholders from the management of *Plasma*. As a result, the participants have two choices. First, they receive dividends as owners from part of the revenue. Second, they work as wage labourers for the company while receiving dividends from their *Plasma* smallholdings. This means that the NES scheme regards participant smallholders partly as land renter and partly as labours. A question therefore arises about the transparency of benefit-sharing and the adaptability of the participating farmers to this scheme. Thus, as with the Paser district, it is necessary to apply various participatory schemes, such as the flexible PMU type and partial schemes, to protect the independence of smallholders.

Regarding land expropriation, the ratio of the nucleus estate and *Plasma* smallholdings is recognised as 80:20, based on the Decree of the Agricultural Minister (*Peraturan Mentri pertanian*) No. 26 2007 and the Decree of the Agricultural Minister No. 98 2013[3], so that companies are allowed to expand nucleus estates broadly. As this chapter has already noted, the possibility of returning the land that was expropriated by companies to villages is very low. Thus, villagers must carefully consider whether they should grow oil palms. In this case, the amount of customary land that would be provided to the company is key. If land expropriation is limited, traditional farms and customary forests could be maintained, such as the case in T village.

'Moderate Industrialisation Based on Rubber Smallholdings with a PMU Scheme'

'Moderate industrialisation' (Kawai and Inoue 2010; Kawai 2011)is a strategy for the partial adaptation of industrialisation and simultaneously maintenance of traditional livelihoods and customary forests of local people, which would be realised by establishing small, dispersed but modern high-yield rubber blocks though a PMU type scheme without land expropriation by a company. Although, as in the case of N village, high-yield rubber smallholdings also may accelerate monoculture development, the development scale and speed are smaller and slower than company's estate development[4]. In the era of globalisation, with rapid environmental destruction and socioeconomic change, 'gradual,' 'well thought out' and 'unhurried' development has significant meaning. Thus, the concept of 'moderate industrialisation' has the potential to address the problems resulting from current oil palm development.

---

[2] Overall, 2 million ha (1.5 million ha of oil palms, 300,000 ha of rubber trees and 200,000 ha of cacao trees (*Theobroma cacao*) of expansion, replanting and rehabilitation was planned (Departemen Pertanian 2007).

[3] The Decree of the Agricultural Minister No. 98 2013 substitutes for the Decree of the Agricultural Minister No. 26 2007.

[4] How control people's rational economic incentives is out of the scope of this study. Combination of moderate industrialization and some forest conservation initiatives such as Payment for Ecosystem Services (PES) and Reducing Emissions from Deforestation and Forest Degradation plus (REDD+) may be one of the solutions to conserve the remaining customary forest.

Figure 3 shows the options for the local people. The options that are chosen depend on the background of each region. Regarding land in critical condition — in which forests are already degraded, *alang-alang* has spread, and no one (including local people) can use these lands — industrialisation through oil palms may bring economic prosperity and efficient land use. In contrast, in regions where there is still significant amounts of primary and secondary forest — and thus many forest products that are utilised by local people still exist — 'moderate industrialisation' could be a strong option from economic, social and environmental points of view.

Address the Challenge of PMU Scheme

In the estate revitalisation program in the West Kutai district, 972 ha of the rubber project application had been accepted by the district estate office as of 2008. A participant bank, BRI (*Bank Rakyat Indonesia*), however, did not accept this application. The main reason was the change of financial mechanisms in estate revitalisation program. The framework of the rubber project without the involvement of the companies in the estate revitalisation program is based on the PMU type scheme[5]. However, there is a major change in financial mechanism. As mentioned before, until the 1990s, the funds of most PMU type projects were supported by government and international development banks. The role of the participant banks at that time was merely to channel the distributed funds to farmers, and the central government had the responsibility to cover defaults. While this effort succeeded in establishing high-yield smallholdings, as this chapter showed, the low repayment rate was an issue. This problem is typical of subsidised and supply-led traditional agricultural credit (Egaitsu 1988; Izumida 2003). In the estate revitalization program, however, support from the central government only takes the form of subsidies that covering part of the credit interest so that participant banks need to use their own funds and provide credit on a commercial basis and have responsibility for the defaults. In general, however, rural financial markets in many developing countries have not matured because of high law-enforcement costs and information asymmetry (Hoff and Stiglitz, 1990). In addition, rural finance generally targets small farmers and small enterprises, which are usually low-income households. This leads to heavy transaction costs. Commercial financial institutions have traditionally avoided serving poor people.

To cope with these problems, some implications can be derived from the case study of this paper. Reducing credit size is an efficient way to improve the repayment incentive of farmers. The cost of the rubber project was 25.7 million rupiah per hectare for five years as of 2008, including labour wages, leading to high risk of default. As shown in the Paser district case, smallholders need not maximise their smallholding productivity. In this case, the small credit size gives participants a repayment incentive. A West Kutai district estate office staff member who had worked on the TCSSP also revealed that the total credit amount for high-yielding rubber smallholdings could be reduced by 10 million rupiah per hectare per five years by using cloned high-yield rubber seedlings and suitable amounts of fertiliser and agrichemicals and that further labour wages need not be paid as beneficiaries/smallholders themselves could serve as both managers and workers. In fact, the PMU type scheme in the

[5] In the estate revitalisation program, however, project management unit is not organized because local estate office in district manage the program. So this scheme does not use the term PMU (*UPP* in Indonesian language). The guidance of the ministry of agriculture called this programme as "Program Revitalisasi Perkebunan tanpa Mitra Usaha (estate revitalization programme without business partners)" (Departemen Pertanian 2007).

Paser district was conducted by this way. Because the amount of credit was small, it was not a burden for participants and most of participants interviewed could complete the credit repayment.

In addition, some Micro Finance Institutions (MFIs) succeeded in achieving sustainable rural finance as part of their social goal to alleviate poverty in recent years. In East Kalimantan, six Credit Unions, semi-formal financial institutions, have been developed by the Dayak people in rural areas, including the Middle-Upper Mahakam region since 1999 (Kawai and Inoue 2011). CUs are developed based on 'common bonds,' such as ethnic, religious, work and community relations and address high transaction costs and information asymmetries in rural areas. A CU has the following two major strong points. First, the CU has village-level repayment schemes, while other formal financial institutions do not. Second, the CU accumulates the loan funds from local people as savings. In the case of research CU, *Credit Union Petemai Urip*, all loans are funded by members' savings. The average deposit balance per depositor and the average loan balance per borrower as of 2008 were 3.96 and 4.77 million rupiah, respectively (Kawai and Inoue 2011, p. 164), which means that this CU mobilises and utilises local financial capital and does not depend on outside funds. If the PMU scheme can reduce the credit size and find optimal input, collaboration and the exchange of skills and knowledge with such local initiatives may contribute to revising the financial mechanism of the PMU scheme. However, sometimes CUs and the local government have different political interests. In such a case, it is not easy to create such a relationship.

For the time being, the government may need to cover costs, such as the initial cost for obtaining core elements, such as high-yielding seedlings and the certificate of land title, until the modified sustainable financial scheme is established, though the total coverage depends on the budget and is limited.

Current Condition

Stagnation of National Estate Revitalisation Program

Because of the challenges in the financial mechanism, the estate revitalization program did not proceed well. Although the program was extended to 2014, the areas that obtained agreement from the banks[6] only totalled 237,315 ha (225,183 ha for oil palm, 10,612 ha for rubber and 1.520 ha for cacao) as of 2014 (Media Perkebunan 2014). The reason that the oil palm development that applied the NES-type scheme proceeded better (though its achievement is quite low from initial plan) than the rubber and cacao developments without the involvement of the companies may be that sometimes nucleus estate companies work as guarantors and cover the initial costs. Because of this stagnation, the ministry of finance decided to suspend new payments for the estate revitalization program in January 2015[7] until a new financial scheme is established (Info Sawit 2015).

---

[6] The name of banks are PT Bank Rakyat Indonesia, PT Bank Mandiri, PT BUKOPIN, PT BNI, PT BPD Sumatera Utara, BPD Sumatera Selatan, BPD Sumatera Barat/Bank Nagari, BPD Riau, BPD NAD, BPD Papua, PT Bank Niaga, PT Bank Agro, Bank Mega, Bank Artha Graha, PT BII, and BPD Kalimantan Timur (Kementerian Pertanian 2013).

[7] According to Surat Menteri Keuangan No. S-5/MK.05/2015.

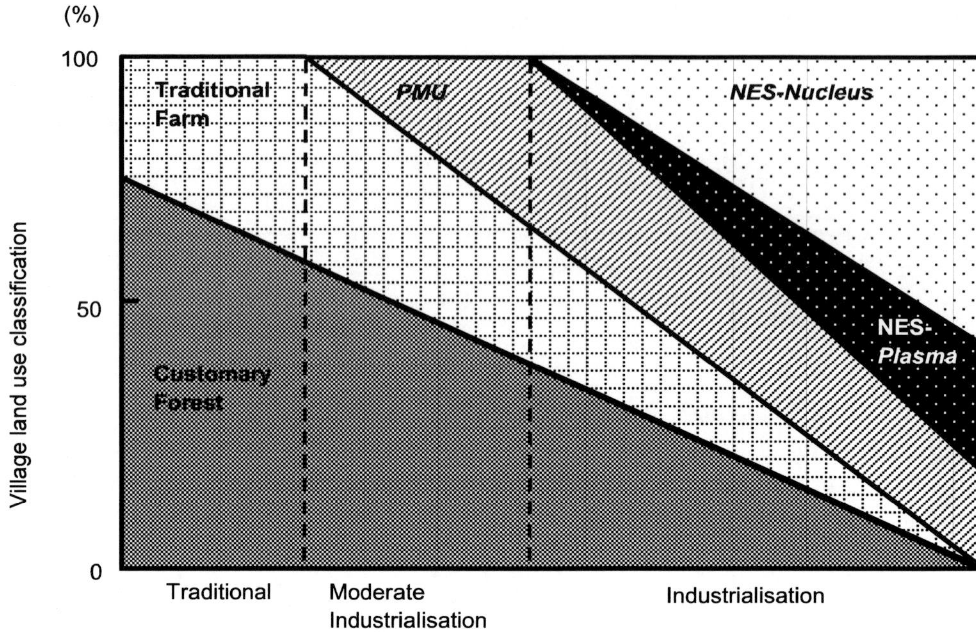

1) : Customary Forest of local people
: Traditional Farm such as swidden, fruit gardens, Arbizia gardens, and traditional rubber gardens
: PMU high-yielding rubber or oil palm smallholdings
: *Plasma* high-yielding Oil Palm Smallholdings

: *NES-Nucleus* intensive high-yielding oil palm plantation.

2) The vertical axis shows village land use classification, and the horizontal axis shows the conversion of customary forests and traditional farms into intensive plantations. The leftmost side shows the recent land use classification of the Middle-Upper *Mahakam* region. The rightmost shows the condition that all village area is converted into intensive plantation by NES and PMU schemes.

Figure 3. Conversion of customary forest and traditional farms of local people into intensive high-yielding plantations and smallholdings (Modified from Kawai 2011).

Oil Palm Development in East Kalimantan

The provincial government of East Kalimantan strongly encourages the palm oil industry and aims to develop one million hectares of oil palm plantations by 2013. Table 6 shows the change of oil palm plantations in East Kalimantan. As of 2007, the total area of oil palm plantations in East Kalimantan was still 339,292 ha. The total area then reached 1.1 million ha in 2013.

In the Paser district, the area of state-owned companies was 22,342 ha, the area of private companies was 88,437ha, and the area of smallholders is 71,167 ha as of 2013. The increase of private companies and smallholder plantations is notable. As this chapter showed, oil palms are accepted by both of migrants and local people, so it can be assumed that the area of smallholders' oil palm increased. However, it is also necessary to pay attention to how the rapid increase of private companies' oil palms has been realized with the possibility of additional land expropriation.

In the West Kutai district, the area of the private companies was 78,741 ha and the area of smallholder was 3,929 ha. Ninety-five percent of oil palm land comprises plantations of private companies. These data indicated that land expropriation is proceeding and the establishment of *Plasma* smallholdings are delayed. In both cases in this chapter, the development of oil palm smallholdings in the Paser district and the development of high-yielding rubber smallholdings in the West Kutai district required 20-30 years for the local people to adapt to new crops and technologies. However, the current oil palm development does not allow the local people time for adaptation. Obviously, one of the reasons is the lack of effective smallholder support schemes and alternative development options in this region.

It is urgently necessary to develop economically attractive but ecologically and socially sound smallholder support institutions that can alleviate the strong expansion pressure of company-led large-scale oil palm monoculture and marginalization of local people through land expropriation.

Table. 6. Area change of oil palm plantation in East Kalimantan, Indonasia (ha)

District/City	2007				2013			
	State-owned company	Private company	Smallholder	**Total**	State-owned company	Private company	Smallholder	**Total**
Paser	13,526	14,798	37,795	**66,119**	22,342	88,437	71,167	**181,946**
West Kutai	0	5,371	100	**5,471**	0	78,741	3,929	**82,670**
North Pasir Penajam	0	18,087	5,988	**24,075**	0	32,040	16,659	**48,699**
Kutai Kartanegara	25	54,340	6,495	**60,859**	25	158,295	20,544	**178,864**
East Kutai	0	83,624	10,360	**93,984**	0	283,377	76,142	**359,519**
Berau	0	28,544	9	**28,553**	0	72,539	19,052	**91,591**
Bulungan	0	8,822	1,604	**10,426**	0	58,290	1,533	**59,823**
Malinau	0	0	0	**0**	0	0	1,050	**1,050**
Nunukan	0	41,395	8,080	**49,475**	0	77,199	18,592	**95,791**
Samarinda	0	0	331	**332**	0	0	1,489	**1,489**
Tana Tidung	0	0	0	**N.A**	0	13,844	81	**13,925**
Balikpapan	0	0	0	**0**	0	0	10	**10**
Bontang	0	0	0	**0**	0	20	18	**38**
Total	13,551	254,981	70,762	**339,294**	22,367	862,782	230,266	**1,115,415**

Source: Statistical data of the East Kalimantan estate office in 2008 and statistical dat on the website for the East Kalimantan estate office: http://disbun.kaltimprov.go.id/statis-35-komoditi-kelapa-sawit.html (Accessed in 26 April 2015)

ACKNOWLEDGEMENTS

This research was carried out under the 'Agreement on Academic Exchange between the Graduate School of Agricultural and Life Sciences, the University of Tokyo and the Mulawarman University, East Kalimantan Indonesia.' This research was supported by the [Japan Society for The Promotion of Science] under the [the Biodiversity and Ecosystem

Restoration (leader: WASHITANI, Izumi, Prof. of The University of Tokyo), which was among the 21st Century Programs at The University of Tokyo]; the [Japan Society for The Promotion of Science] under Grants-in-Aid for Scientific Research (B) [Relationships among stakeholders for Tropical Satoyama governance (leader: ICHIKAWA, Masahiro, Prof. of Kochi University, No. 2040101)]; the [Japan Society for The Promotion of Science] under the Grant Aid for Scientific Research on Priority Areas in Multi-level Environmental Governance for Sustainable Development [Management of Local Commons in the Global Era (leader: MUROTA, Takeshi, Prof. of Doshisha University, No. 18078009)]; the [Japan Society for The Promotion of Science] under Grants-in-Aid for Scientific Research (A) [Oil palm smallholders and estate companies: the behaviour changes with globalization and the impact (leader: HAYASHIDA, Hideki, Associate Prof. of Doshisya University, No. 26257005)].

REFERENCES

Asian Development Bank (1997). *Impact Evaluation Study of Bank Operation in the Industrial Crops and Agro-Industry Sector in Indonesia,* Manila.

Asian Development Bank (2002). "Project Completion Report on the Tree Crop Smallholder Sector Project (Loan 118-INO) in Indonesia, Manila.

Asian Development Bank (2004a). Project Completion Report On The Upland Farmer Development Project (Loan 1184-INO[SF]) In Indonesia, Manila.

Asian Development Bank (2004b). Project Completion Report On The Sulawesi Rain-Fed Agriculture Development Project (Loan 1351-INO[SF]) In Indonesia, Manila.

Asian Development Bank (2005). Project Completion Report, Small holder Tree Crop Processing Project (Loan 1186-INO) in Indonesia, Manila.

Anne, C., (2000). The Hesitant Boom: Indonesia's Oil Palm Sub-Sector in an Era of Economic Crisis and Political Change Center for Institutional Forest Research (Center for International Forest Research), Bogor.

Barlow, C. and Jayasuriya, S. K., (1984). Problems of Investment for Technological Advance: The Case of Indonesian Rubber Smallholders. *Journal of Agricultural Economics*, 35 (1): 85-95.

Barlow, C. and Tomichi, T., (1991). Indonesian Agricultural Development: The Awkward Case of Smallholder Tree Crops. Bulletin of Indonesian Economic Studies, 27 (3): 29-53.

Boeke J. H. 1953. Economics and Economic Policy of Dual Societies as exemplified by Indonesia: International secretariat institute of pacific Relations. New York., Trans. Nagai. K. (1971): Syukin Shobo. (Japanese)

Carlson K. M, Curran L. M, Asner G. P, Pittman A. M, Trigg S. N, Adeney J. M., (2013). Carbon emissions from forest conversion by Kalimantan oil palm plantations. *Nature climate change* 3: 283-287.

Douglas S., Anne, C., Erik, M., Meine, N., Joanne, G., Jasqui, S-G., Karah, W. and Markku, K., (2009). The Impacts and Opportunities of Oil Palm in Southeast Asia, What do we know and what we need to know?, Occasional Paper No. 51, Center for International Forest Research, Bogor.

Departemen Pertanian, (1984). Rencana Pembangunan Lima Tahun Keempat Sub Sektor Perkebunan Pola UPP II, Departmen Petanian Direktorat Jenderal Perkebunan, Jakarta.

Departemen Pertanian, (1985). Laporan Pelaksanaan Proyek Peremajaan, Rehabilitasi Dan Perluasan Tanaman Ekspor (PRPTE) Selama Pelita III (1979/1980 s/d 1983/1984), Buku-II, Jakarta.

Departemen Pertanian, (1986). Hasil Pertemuan Konsultasi dan Pembinaan UPP Proyek PRPTE Cipayung, 9 s/d 19 Desember 1985 BUKU I, Hasil Perumusan dan Makalah Direktorat Jenderal Perkebunan, Jakarta.

Departemen Pertanian, (1992) Perusahaan Inti Rakyat Perkebunan: Pelaksanaan Dan Penilaian. Direktorat Jenderal Perkebunan Tim Khusus Proyek Perkebunan Inti Rakyat. Jakarta. 1 Nopember 1992

Departemen Pertanian, (1999). Laporan Perkembangan Pelaksanaan Proyek-Proyek PIR (PIRBUN, PIR TRANS, PIR BUN KKPA & PIR TRANS KKPA), Direktorat Jenderal Perkebunan Sekretariat PIR, Jakarta.

Departemen Pertanian (2006). Laporan Perkembangan, Eks. Proyek-Proyek Pola UPP Perkebunan, Direktorat Jenderal Perkebunan, Departmen Pertanian, Jakarta.

Departemen Pertanian (2007). Pedoman Umum, Program Revitalisasi Perkebunan (Kelapa Sawit, Karet dan Kakao), Direktorat Jenderal Perkebunan, Jakarta.

Egaitsu, F. 1988. Rural Financial Markets in Asia: Two Schools of Thought. In: *Farm Finance and Agricultural Development:* 111-122. Asian Productivity Organization. Tokyo.

Feintrenie L, Chong W. K, Levang L., (2010). Why do Farmers prefer oil palm? Lessons Learnt from Bungo District, Indonesia. *Small-scale Forestry* 9 (3): 379-396.

Gillespie P., (2012). Participation and power in Indonesian oil palm plantations. *Asia Pacific Viewpoint* 53 (3): 254–271.

Hoff, K. and Stiglitz, J. E., (1990). Introduction: Imperfect Information and Rural Credit Markets - Puzzles and Policy Perspectives. *The World Bank Economic Review* 4 (3): 235-250.

Info Sawit (2015). Moratorium Revitbun Pembiayaan Petani Terancam Terhenti, Info Sawit. http://www.infosawit.com/mobile/index.php/news/detail/moratorium-revitbun--pembiayaan-petani-terancam-terhenti (Accessed in 12 Oct. 2015).

Inoue, M., (1994). Forest Utilization and Economic Development in Indonesia, Nagata, S., Ionue M., Oka H., "Forest Resource Utilization and Regeneration: Economic logic and Natural Logic," Rural Culture Association Japan: 92-145 (Japanese).

Izumida, Y. (2003). *Rural Development Finance: Asian Experience and Economic Growth.* University of Tokyo Press, 221 pp. (In Japanese).

Kawai M. and Inoue M., (2010). Possibility of Moderate Industrialization as Alternative Development Strategy to Oil Palm Plantation: The case of Middle-Upper Mahakam Region in East Kalimantan, Indonesia. *Forest Economy* 63(7): 1-17. (in Japanese).

Kawai M, (2011). "Moderate Industrialization" and Commons: Alternative Development Strategy to Oil Palm Plantation in East Kalimantan, Indonesia. Paper presented at: 13th Biennial Conference of the International Association for the Study of the Commons (IASC); Hyderabad. Available from: http://iasc2011.fes.org.in/papers/docs/432/submission/original/432.pdf.

Kawai M. and Inoue M., (2011). The possibility and challenge of achieving sustainable rural finance by a credit union of the Dayak people: The case of the middle-upper Mahakam region in East Kalimantan, Indonesia, *TROPICS*, Vol. 19 (4): 157-178.

Kementerian Pertanian (2013), Peningkatan Produksi, Produktivitas dan Mutu Tanaman Tahunan, Pedoman Teknis Revitalisasi Perkebunan (Kelapa Sawit, Kakao, Karet) Tahun 2014, Direktorat Jenderal Perkebunan, Kementerian Pertanian Desember 2013.

Koh L. P. and Wilcove D. S., (2008). Is Oil palm agriculture really destroying tropical biodiversity? *Conservation Letters* 1 (2): 60-64.

Marti S., (2008). *Losing ground: the human rights impacts of oil palm plantation expansion in Indonesia.* Dilworth A, Baird N, Kirby J, editor. Friends of the Earth, LifeMosaic and Sawit Watch.

McCarthy J. F., (2010). Processes of inclusion and adverse incorporation: oil pal and agrarian change in Sumatra, Indonesia. *The Journal of Peasant Studies* 37 (4): 821-850.

McCarthy J. F., Gillespie P., Zen Z., (2012). Swimming Upstream: Local Indonesian Production Networks in "Globalized" Palm Oil Production. *World Development* 40 (3): 555–569.

Media Perkebunan, (2014). Realisasi Revitalisasi Perkebunan, Media Perkebunan. http://www.mediaperkebunan.net/index.php?option=com_content&view=article&id =659%3Arealisasi-revitalisasi-perkebunan& catid=17%3Aberita&Itemid=24 (Accessed in 29 June 2015).

Ministry of Forestry and Estate Crops, (2000). Tree Crops Smallholder Development Project (TCSDP) IBRD LOAN NO. 3464-IND, Final Evaluation Report on TCSDP, Ministry of Forestry and Estate Crops, Directorate General of Estate Crops, Jakarta.

Rist L., Feintrenie L., Levang P., (2010). The livelihood impacts of oil palm: smallholders Indonesia, *Biodiversity and Conservation* 19 (4): 1009-1024.

Wilcove D. S., and Koh L. P., (2010). Adressing the threats to biodiversity from oil-palm agriculture. *Biodivers Conserv*, 19 (4): 999-1007.

Zen, Z., Barlow, C., and Gondowarsito, R. (2006). "Oil palm in Indonesia socio-economic improvement: a review of option." *Oil Palm Industry Economic Journal* 6 (1): 18-29.

In: Monoculture Farming
Editors: Tapan Kumar Nath and Patrick O'Reilly

ISBN: 978-1-63485-166-4

Chapter 7

ARBUSCULAR MYCORRHIZAL FUNGI UNDER MONOCULTURE FARMING: A REVIEW

***Marcela Claudia Pagano*[1,*] *and Partha P. Dhar*[2]**
[1]Department of Physics, Federal University of Minas Gerais, Belo Horizonte, Minas Gerais, Brazil
[2]Plant Production Dept, College of Food and Agricultural Sciences, King Saud University, Riyadh, Saudi Arabia

ABSTRACT

The actual tendency of conventional agriculture to grow crops in monoculture to meet the demands of both bioenergy and food production can result in crop yield decline or soil degradation and increased use of synthetic fertilizers. Rhizospheric and soil microorganisms can play a crucial role in plant growth and plant health, contributing to agroecosystem services in many ways. Most crops and terrestrial plants associated with mycorrhizas can lead to better plant growth under stressful conditions through symbiosis. Various arbuscular mycorrhizal fungi (AMF) structures can be used to estimate AMF abundance and diversity in crop fields. The abundance and diversity of AMF are expected to be lower in monocultures. With the increasing effects of global change on field crops, the associated plant-soil microbial communities have been more studied and their beneficial function more recognized for plant success. The lack of long term field studies fail in giving results of the complex cropping systems. Better knowledge of the wide variation in AMF ecology is important to understand and better manage agroecosystems. This chapter explores the current available information relevant to benefits/drawbacks of AMF symbioses in monocultures. Possible roles of fungal structures and Glomalin are also discussed.

Keywords: plant monoculture, mycorrhizal infection rate, mycorrhizal spores, glomalin

* Corresponding author address: Dr Marcela Claudia Pagano, Brazil. Email: marpagano@gmail.com; mcp2006@ufmg.br.

INTRODUCTION

Monoculture usually refers to single species vegetation. However, cultivation of a single crop over a large area for consecutive years is the agronomic definition of monoculture (Miller and Spoolman 2009). It is widely practiced in the modern commercial and industrial agricultural system with the aim of higher production in the agricultural sector as well as industrial sector. Monoculture is practiced in the agricultural systems for commercial food products, animal feed, oil, agro-fuels; in the forestry section for paper industries, charcoal and timber products (Evans 2009).

In the monoculture system, cultivated crops usually are of identical genetic similarity, same growth patterns and resistant to certain common disease as such. Crops cultivated in the monoculture farming are alike to habits and high yielding quality; and their minimum land requirement, maintenance and managements of the crop species can easily be homogenized. Monoculture cultivation includes the crop species exclusively suitable to the specific situation of a particular location, e.g., salinity, drought, a diminutive growing season. Homogenous management of monoculture farming facilitates the crop production reducing the waste and loss from inefficient transplantation and harvesting. Due to similar genetic constituents of the crop plants used in the monoculture system, less competition exists between the plant species for nutrient resources, space and solar energy. In a word, high profit is pulled out from the yielding of crops cultivated in the monoculture farming. It is the ideal in the largest commercial agriculture and in terms of economies of scale; it automates the production process and thus reducing the costs.

Instead of commercial benefits, monoculture farming exerts destructive consequences in the long run. Genetically narrowness of the monoculture crops challenges the adaptability of the crops in diverse environmental situations or diverse climatic circumstances. Moreover, susceptibility to the pathogenic causal organisms and disease incidence are increased. Besides, decline in soil microbial diversity, providence of narrow ranged-habitats and reducing the composition and abundance of biodiversity as such, increased nutrient deficiency as well as decreased soil fertility are all the impacts of monoculture farming. On the other hand, increased usage of chemical inputs exerts deleterious effects on the soil ecosystem by increasing soil pollution and negative environmental impacts. Improper usage of the toxic chemicals causes the different physiological disorder including carcinogenic symptoms as well. In a word, modern monoculture destroys the wild ecosystem (Altieri 1992).

Most plant species associated with microorganisms in a mutually beneficial way, for example, legume plants associate with rhizobial bacteria, and most plants with mycorrhizas and/or endophytic microorganisms. Associations with *Trichoderma* and dark septate fungi are increasingly studied. However, mycorrhizal interactions may have important functions in terrestrial ecosystems, being a topic of accelerated developing research. The arbuscular mycorrhizal fungi (AMF) are being specially prospected due to their multiple functions in terrestrial ecosystems. These plant-soil symbionts are able to perform better under stressful conditions and the increasing recognition of their importance has led to an increasing number of research groups (mostly in Ecology sciences) worldwide. The AMF structures such as total root colonized, percent root length colonized by arbuscules can be used to estimate of AMF abundance within a plant or ecosystem. The extra radical hyphae of AMF extend from roots into the surrounding soil much further than do roots and establish the mycorrhizosphere

(Azcón 2014). The study of AMF in different agro-ecosystems can contribute to better understand the effect of global change on plants and soil associated communities. Many crops associate with AMF and this symbiosis has to be monitored under global change. It is known that land-use intensity will increase in the future for both food and bioenergy production, and long rotations are seen as not practical (reviewed by Bennet et al., 2012). With the increasing global climatic changes an increase of temperature (2–7 °C) is expected, and also changes in distribution of precipitations and in storm intensity (reviewed by Wu et al., 2011). The anthropogenic influence with release of greenhouse gases on climate was recognized, as well as the fact that several agricultural activities release CH_4 and N_2O. Thus, agriculture is a major contributor to global and climate changes and the presence of AMF in soils may be crucial for sustainable agriculture. The AMF are natural "biofertilizers" and they can contribute to the management of fertilizers as well as other environmental constraints decreasing ecological consequences (Withers et al., 2014). Better knowledge of the variation in mycorrhizas is important for management of monocultures in different regions.

THE AMF IN CROPS

Most soils contain native AMF that colonize crop roots promoting growth and phosphorus (P) uptake of plants, which can benefit of inoculation with AMF. Nowadays, reduced use of agrochemicals and fertilizers together with search for crop sustainability leaded to the promising approach of obtaining high yields with low fertilizer inputs in order to support sustainable agricultural systems (See Siddiqui et al., 2008).

With the increasing effects of global change on field crops, the associated plant-soil microbial communities have been more studied and their function more recognized (Table 1). Additionally, research on AMF comparing tillage systems (conventional tillage and direct seeding) in culture rotations is frequently reported as by Schalamuk et al., (2006) in Argentina, Brito et al., (2012) in Portugal, to name just a few (See Pagano 2011).

Table 1. Journal articles dealing with arbuscular mycorrhizas in most studied monocultures

Key words	Number of journal articles
Monoculture + AMF	58
Citrus + AMF	48
Grapevine + AMF	20
Olive + AMF	19
Maize monoculture + AMF	11
Monoculture trees + AMF	9

Database survey conducted in January 2015 (SCOPUS).

AM FUNGI IN THE MONOCULTURE FARMING

In a monoculture system, AMF proliferated selectively. Mycorrhizal fungi were less beneficial or even detrimental to the crops in which they proliferated (Johnson et al., 1992).

Through hyphal common mycorrhizal networks, mycorrhizal fungi may influence the distribution of limiting mineral nutrients among interconnected individuals, potentially affecting competition and size inequality of the individual in the monocultures (Weremijewicz and Janos, 2013). A potential decline of monocultures contributed significantly to the positive net biodiversity effect over time (Marquard et al., 2013). Long periods of monoculture often reduce soil quality, microbial diversity and community structure. Intercropping with inoculation with the AMF can improve soil quality and plant performance than continuous monoculture (Sheng et al., 2012). Seasonal soil microbial activity could be differentiated under rotational and monoculture management (Marais et al., 2012). In comparison with the monoculture, mixed cropping system provides the more incidence of mycorrhiza in the deeper soil layers. Abundant mycorrhizal incidence at deeper soil layers may contribute to the dynamics of phosphorus cycling in the soil and making nutrients more available to the plant species (Cardoso et al., 2003). The diversity of AMF influences the plant-diversity, productivity, and the stability, sustainability, and functioning of the ecosystem. The monoculture plantation of single species negatively affects the natural AMF diversity in the soil and results in an ecological degraded ecosystem with low biodiversity, including low diversity of AMF species (Sharmah and Jha, 2014).

The AMF symbiosis in maize was more studied in several countries worldwide (Pagano and Covacevich, 2011) and several research groups dedicated to understand the symbioses in maize (Miransari et al., 2007). In Italy, Viti et al., (2010) showed that application of compost (selected urban waste) influence AMF composition. Also in Latino-America important research was achieved for the selection of efficient phosphate solubilizers microorganisms (bacteria and fungi) to be applied in the crop (Gomes et al., 2014; Pagano and Covacevich, 2011; Bedini et al., 2007). Most field studies have focused on top-soil (surface soil); however, Oehl et al., (2005) suggested the research in the deep soil layers such as 50–70 cm depth of vineyard and mono-cropped maize fields, in order to know better the AMF diversity.

It is known that crop establishment has led to loss of soil C stocks (Guo et al., 2002), rapid soil erosion (Prasuhn, 2012). Soil organic C, soil macro-aggregates, catabolic diversity and microbial biomass also declined under cotton monocultures (Acosta-Martínez et al., 2010) and under persistent cultivation of maize (Kösters et al., 2013; Lauer et al., 2011). It is also known that extended cultivation of arable crops and tillage can change the microbial communities towards predominance of Gram-positive bacteria (Tian et al., 2012) and the fungal-to-bacterial biomass ratio decrease (Bailey et al., 2002).

It is reported that AMF prevail in the colonization of maize roots and maize is frequently utilized to propagate AMF in greenhouse and fields. Higher root colonization by AMF can be found in plots fields cultivated with beans and previously cultivated with maize. Moreover, higher root colonization and number of propagules can be found in non-tillage systems than in conventional systems or when the soil is minimally disturbed (Gosling et al., 2006). However, it depends of plant, fungal species, soil type, and environmental conditions. For example, Johnson et al., (1992) found different AMF in maize than in soy fields (monocultures) and also that AMF associated with maize promoted soy growth and vice-versa.

Among industrial crops, maize is very susceptible to soil compaction (Maddonni et al., 1999; Amato and Ritchie, 2002). In China high levels of P and a poor soil structure are detrimental to mycorrhizas (Xiang et al., 2014). In Brazil, Oliveira et al., (2009) showed that colonization by AMF can affect maize productivity (different AMF species were found in the rhizosphere of efficient genotypes in low P level conditions). Boomsma and Vyn (2008) recompiled benefits of AMF in maize production systems, pointing to promote AMF colonization and survival, as well to understand effects of AMF in morpho-physiology of this crop. They stressed that use of AMF is applicable even in intensively cultivated soils.

In order to select inoculants for crops, soil characteristics must be noted. In this regard, in experiments with maize Imaz et al., (2014) observed that AMF native to soils closest to the Province of Buenos Aires were found more efficient to compensate for the deficiencies of Zn, which may be due to the greater AMF diversity.

In Mexico, the AMF symbioses in native maize landraces (cultivated as part of traditional polyculture systems) which are being substituted by fertilized monocultures of genetically modified plants were investigated (also at different level of phosphorus fertilization) (Sangabriel-Conde et al., 2015). The diversity found was high, most AMF species being colonized several maize varieties and a moderate P level was better. The authors pointed the reduction of AMF richness with replacement of these landraces. Interestingly, the Black variety, presented the highest AMF colonization and P uptake, and AMF richness. Three-year monocultures of mycorrhizal medicinal plants species showed that the choice of plant species in a monoculture, can substantially alter the AMF potential of soils which in turn can change the performance of other medicinal plants cultivated subsequently (Zubek et al., 2013).

In agrosystems, factors such as host crop dependency to AMF colonization, tillage system, fertilizer application, and the potential of AMF inoculation, play a role in plant benefits from these fungi and increasing interest in the propagation of AMF for sustainable agriculture is seen (See Siddiqui et al., 2008). The status of the Brazilian inventory of glomalean fungi in agroecosystems has been reviewed (Stürmer and Siqueira, 2006), coffee systems showing up as the best studied crop; however a high proportion of undescribed AMF species was recorded in the studied systems.

Table 2. Selected book, book chapters, reviews or papers dealing with AMF in monocultures. * Review

Ecosystem type/Focus	Country	References
Crop lands AMF	Argentina and Brazil	Pagano and Covacevich (2011)*
Sustainable agriculture, forestry mycorrhizas	Several countries	Siddiqui et al., (2008)*
Native trees mycorrhizas	Brazil	Pagano et al., (2010)*
Corn, wheat	Iran	Miransari et al., (2007, 2009a.b)
Crop lands Soybean, cassava AMF	Brazil	Miranda (2008)
Crop lands Mycorrhizas	Several countries	Smith and Read (2008)
Agroecosystems	Brazil	Stürmer and Siqueira (2006)

It is known that AMF contributes for soil C stock and aggregate stability, through physical and chemical actions. The mycelium network formed and a soil protein (glomalin) secreted by AMF (see Rillig et al., 2001; Rillig, 2004) improve soil quality. However, glomalin responds to land-use change (Rillig et al., 2003) and a positive correlation between glomalin-related soil protein (GRSP) and AMF parameters (spore density and bio-volume) can be obtained (Bedini et al., 2007). Previous reports on higher contents of glomalin in soils from organic than from conventional agriculture in Korea (Kil et al., 2009) point to the use of glomalin as a useful indicator for the assessment of soil health in sustainable agriculture (See Pagano and Covacevich 2011, Table 2). The olive production in Europe is also the objective of new technologies for their cultivation including associated mycorrhizas as inoculants (see Bompadre et al., 2014).

Natural Plots Used as Reference

Among soil ecologists it is usually and convenient to compare the agricultural field microbiota with a native adjacent area. Mycorrhizal interactions between plants, fungi, and the environment are complex and occur often attached (Siddiqui et al., 2008), which is difficult to understand the field results. In general, grassland adjacent to field experiments need to be integrated to the studies of AMF in parallel, in order to cover the largest scale possible of soil exploitation intensity of the same soil type (Oehl et al., 2003, 2009).

Xiang et al., (2014) showed the impacts of maize culture in decreasing AMF richness and their structures (extra-radical hyphae) with regard to native grasslands. Bedini et al., (2007) observed that maize monoculture, grassland and poplar grove – maize soil showed low AMF spore number. A good correlation between glomalin-related soil protein and spore bio-volume suggested that glomalin-related soil protein may represent a useful biochemical parameter for the assessment of biological soil fertility in sustainable agriculture.

Hijri et al., (2006) showed that communities of AMF in continuous maize monoculture can decrease in diversity, but high diversity of AMF was found in long term field experiments where low-input agriculture involving crop rotation provided better conditions to preserve AMF.

Murugan et al., (2014) studied the effects of permanent grassland (100% ryegrass—PG), conversion to modified grassland (mixture of grass and clover—MG) or maize monoculture (MM) on the dynamics of microbial biomass, fungal ergosterol and microbial residues in temperate grasslands of Germany. Very limited information is available on changes in fungal and bacterial residues and microbial functional diversity during conversion of permanent grassland to monoculture (Grayston et al., 2004; Lauer et al., 2011). Mycorrhizal occurrence, parameters and characteristics are of great potential for understanding the ecology of monocultures.

Conclusion

In this chapter, the AMF occurrence and contribution to the agrecosystems based on monocultures are described. The appreciation is that monocultures associated with less and different functional types of mycorrhizas can give different benefits. The combined results are important to understand the effects of AM fungi on crops, and their agricultural and ecological significance. Further studies are required to understand crop microorganisms associations as well as their behavior during agronomical succession. Finally, this chapter showed that monocultures and their associated AMF require urgent research. Preserving the biodiversity of AMF will aid to maintain the ecosystem services even under monocultures.

REFERENCES

Acosta-Martínez, V., Bell, C. E., Morris, B. E. L., Zak, J., Allen, V. G., 2010. Long-term soil microbial community and enzyme activity responses to an integrated cropping–livestock system in a semiarid region. *Agric. Ecosyst. Environ.* 137, 231-240.

Altieri, M. A., 1992. Agroecological foundations of alternative agriculture in California. *Agriculture, Ecosystems and Environment* 39, 23-53.

Amato, M., Ritchie, J. T., 2002. Spatial distribution of roots and water uptake of maize (*Zea mays* L.) as affected by soil structure. *Crop. Sci.* 42, 773-780.

Azcón, R., 2014. *Mycorrhizosphere: The role of PGPR*. In: Morte, A, Varma, A. (eds.) 107-144. Root engineering. Springer, Berlin.

Bailey, V. L., Smith, J. L., Bolton, H., 2002. Fungal-to-bacterial ratios in soils investigated for enhanced C sequestration. *Soil Biol. Biochem.* 34, 997-1007.

Guo, L. B., Gifford, R. M., 2002. Soil carbon stocks and land use change: a meta analysis. *Glob. Change Biol.* 8, 345-360.

Bedini, S., Avio, L., Argese, E., Giovannetti, M., 2007. Effects of long-term land use on arbuscular mycorrhizal fungi and glomalin-related soil protein. *Agriculture, Ecosystems and Environment* 120, 463-466.

Bennett, A. J., Bending, G. D., Chandler, D., Hilton, S., Mills, P., 2012. Meeting the demand for crop production: the challenge of yield decline in crops grown in short rotations. *Biol. Rev.* 87, 52-71.

Bompadre, M. J., Colombo, R., Molina, M. C. R., Godeas, A. M., Pardo, A. G., 2014. Arbuscular Mycorrhizal Fungi in the Alleviation of Oxidative Stress under Cutting Propagation Management (pp. 181-202) In: Mycorrhizas: Structure, Development and Functions, Warwick E. (ed).

Boomsma, C. R., Vyn, T. J., 2008. Maize drought tolerance: Potential improvements through arbuscular mycorrhizal symbiosis? *Field Crops Research*, 108, 14-31.

Brito, I., Goss, M. J., Carvalho, M., Chatagnier, O., van Tuinen, D., 2012. Impact of tillage system on arbuscular mycorrhiza fungal communities in the soil under Mediterranean conditions. *Soil & Tillage Research* 121, 63-67.

Cardoso, I. M., Boddington, C., Janssen, B. H., Oenema, O., Kuyper, T. W., 2003. Distribution of mycorrhizal fungal spores in soils under agroforestry and monocultural coffee systems in Brazil. *Agroforestry Systems* 58, 33-43.

Evans, J., 2009. Planted forests: uses, impacts and sustainability. FAO and CAB International, Wallingford, UK, 213 p.

Gosling, P., Hodge, A., Goodlass, G., Bending, G. D., 2006. Arbuscular mycorrhizal fungi and organic farming. *Agriculture, Ecosystems and Environment* 113, 17-35.

Grayston, S. J., Campbell, C. D., Bardgett, R. D., Mawdsley, J. L., Clegg, C. D., Ritz, K., Griffiths, B. S., Rodwell, J. S., Edwards, S. J., Davies, W. J., Elston, D. J., Millard, P., 2004. Assessing shifts in microbial community structure across a range of grasslands of differing management intensity using CLPP, PLFA and community DNA techniques. *Appl. Soil Ecol.* 25, 63-84.

Hijri, I., Sykorová, Z., Oehl, F., Ineichen, K., Mäder, P., Wiemken, A., Redecker, D., 2006. Communities of arbuscular mycorrhizal fungi in arable soils are not necessarily low in diversity. *Molecular Ecology* 15, 2277-2289.

Imaz, P. A., Barbieri, P. A., Echeverría, H. E., Rozas, H. R. S., Covacevich, F., 2014. Indigenous mycorrhizal fungi from Argentina increase Zn nutrition of maize modulated by Zn fertilization. *Soil Environ.* 33(1), 23-32.

Johnson, N. C., Copeland, P. J., Crookstand, R. K., Pfleger, F. L., 1992. Mycorrhizae: possible explanation for yield decline with continuous corn and soybean. *Agron. J.* 84, 387-390.

Johnson, J. M.-F., Franzluebbers, A. J., Weyers, S. L., Reicosky, D. C., 2007. Agricultural opportunities to mitigate greenhouse gas emissions. *Environmental Pollution* 150, 107-124.

Kil, Y. J., Lee, J. E., Shin, H. C., 2009. *Diversity of arbuscular mycorrhizal fungi and glomalin.* In: 6th International Conference on Mycorrhiza ICOM6, 2009, Belo Horizonte. Abstracts ICOM6, Viçosa editors, 1, 40.

Kösters, R., du Preez, C. C., Preger, A. C., Amelung, W., 2013. Reaggregation dynamics of degraded cropland soils with prolonged secondary pasture management in the South African Highveld. *Geoderma* 192, 173-18.

Lauer, F., Kösters, R., du Preez, C. C., Amelung, W., 2011. Microbial residues as indicators of soil restoration in South African secondary pastures. *Soil Biol. Biochem.* 43, 787-794.

Maddonni, G. A., Urricariet, S., Ghersa, C. M., Lavado, R. S., 1999. Assessing soil quality in the Rolling Pampa (Argentina), using soil properties and maize characteristics. *Agron. J.* 91, 280-287.

Marquard, E., Schmid, B., Roscher, C., De Luca, E., Nadrowski, K., et al., 2013. Changes in the Abundance of Grassland Species in Monocultures versus Mixtures and Their Relation to Biodiversity Effects. *PLoS ONE* 8(9): e75599.

Miller, G. T., Spoolman, S. E., 2009. *Living in the Environment: Principles, Connections, and Solutions.* Cengage learning Inc., 583 p.

Miransari, M., Bahrami, H. A., Rejali, F., Malakouti, M. J., Torabi, H., 2007. Using arbuscular mycorrhiza to reduce the stressful effects of soil compaction on corn (*Zea mays* L.) growth. *Soil Biology & Biochemistry* 39, 8, 2014-2026.

Miransari, M., Bahrami, H. A., Rejali, F., Malakouti, M. J., 2009a. Effects of soil compaction and arbuscular mycorrhiza on corn (*Zea mays* L.) nutrient uptake. *Soil & Tillage Research* 103, 282-290.

Miransari, M., Bahrami, H. A., Rejali, F., Malakouti, M. J., 2009b. Effects of arbuscular mycorrhiza, soil sterilization, and soil compaction on wheat (*Triticum aestivum* L.) nutrients uptake. *Soil & Tillage Research* 104, 48-55.

Marais, A., Hardy, M., Booyse, M., Botha, A., 2012. Effects of Monoculture, Crop Rotation, and Soil Moisture Content on Selected Soil Physicochemical and Microbial Parameters Fields. *Applied and Environmental Soil Science*, Article ID 593623, 13 p. http://dx.doi.org/10.1155/2012/593623

Miranda, J. C. C., 2008. Cerrado, micorriza arbuscular, ocorrência e manejo. Embrapa Cerrados, Planaltina, 169p. (in Portuguese).

Murugan, R., Loges, R., Taube, F., Sradnick, A., Joergensen, R. G., 2014. Changes in Soil Microbial Biomass and Residual Indices as Ecological Indicators of Land Use Change in Temperate Permanent Grassland. *Microb. Ecol.* 67, 907-918.

Oehl, F., Sieverding, E., Ineichen, K., Mäder, P., Boller, T., Wiemken, A., 2003. Impact of land use intensity on the species diversity of arbuscular mycorrhizal fungi in agroecosystems of Central Europe. *Applied and Environmental Microbiology* 69, 2816-2824.

Oehl, F., Sieverding, E., Ineichen, K., Ris, E., Boller, T., Wiemken, A., 2005. Community structure of arbuscular mycorrhizal fungi at different soil depths in extensively and intensively managed agroecosystems. *New Phytologist* 165, 273-283.

Oehl, F., Sieverding, E., Ineichen, K., Mäder, P., Wiemken, A., Boller T., 2009. Distinct sporulation dynamics of arbuscular mycorrhizal fungal communities from different agroecosystems in long-term microcosms. *Agriculture, Ecosystems and Environment* 134, 257-268.

Oehl, F., Jansa, J., Ineichen, K., Mäder, P., van der Heijden. M., 2011. Champignons mycorhiziens arbusculaires, bioindicateurs dans les sols agricoles suisses. *Recherche Agronomique Suisse* 2 (7–8), 304-311.

Oliveira, C. A., Sa, N. M. H., Gomes, E. A., Marriel, I. E., Scotti, M. R., Guimarães, C. T., Schaffert, R. E., Alves, V. M. C., 2009. Assessment of the mycorrhizal community in the rhizosphere of maize (*Zea mays* L.) genotypes contrasting for phosphorus efficiency in the acid savannas of Brazil using denaturing gradient gel electrophoresis (DGGE). *Applied Soil Ecology* 41, 249-258.

Pagano, M. C., Covacevich, F., 2011. Arbuscular Mycorrhizas in Agroecosystems. In: Fulton, S. M., editor. *Mycorrhizal Fungi: Soil, Agriculture and Environmental Implications*. Nova Science Publishers, New York. pp. 35-65.

Pagano, M. C., 2011. Soil tillage in agroforestry and agroecosystems: Mycorrhizal benefits. In: Soil Tillage and Microbial Activities, Editor: M. Miransari.

Prasuhn, V., 2012. On-farm effects of tillage and crops on soil erosion measured over 10 years in Switzerland. *Soil Till. Res.* 120, 137-146.

Rillig, M. C., Wright, S. F., Nichols, K. A., Schmidt, W. F., Torn, M. S., 2001. Large contribution of arbuscular mycorrhizal fungi to soil carbon pools in tropical forest soils. *Plant and Soil* 233, 167-177.

Rillig, M. C., Ramsey, P. W., Morris, S., Paul, E. A., 2003. Glomalin, an arbuscular-mycorrhizal fungal soil protein, responds to land-use change. *Plant and Soil* 253, 293-299.

Rillig, M. C., 2004. Arbuscular mycorrhizae, glomalin, and soil aggregation. *Canadian Journal of Soil Science* 84, 355-363.

Sangabriel-Conde, W., Maldonado-Mendoza, I. E., Mancera-López, M. E., Cordero-Ramírez, J. D., Trejo-Aguilar, D., Negrete-Yankelevich, S., 2015. Glomeromycota associated with Mexican native maize landraces in Los Tuxtlas, Mexico. *Applied Soil Ecology* 87, 63-71.

Sheng, P. P., Liu, R. J., Li, M., 2012. Inoculation with an Arbuscular Mycorrhizal Fungus and intercropping with pepper can improve soil quality and watermelon crop performance in a system previously managed by monoculture. *American-Eurasian J. Agric. Environ. Sci.*, 12, 1462-1468.

Siddiqui, Z. A., Akhtar, M. S., Futai, K., (eds.) 2008. *Mycorrhizae: Sustainable Agriculture and Forestry*. Springer, 350p.

Stürmer, S. L., Siqueira, J. O., 2006. Diversity of arbuscular mycorrhizal fungi in Brazilian ecosystems. In: Moreira, F. M. S., Siqueira, J. O., Brussard, L. (eds.) *Soil biodiversity in Amazonian and other Brazilian ecosystems*. Wallingford: CABI-Publications, pp. 206-236.

Viti, C., Tatti, E., Decorosi, F., Lista, E., Rea, E., Tullio, M., Sparvoli, E., Giovannetti, L., 2010. Compost Effect on Plant Growth-Promoting Rhizobacteria and Mycorrhizal Fungi Population in Maize Cultivations. *Compost Science & Utilization* 18, 4, 273-281.

Weremijewicz, J., Janos, D. P., 2013. Common mycorrhizal networks amplify size inequality in Andropogon gerardii monocultures. *New Phytologist* 198, 203-213.

Withers, P. J. A., Sylvester-Bradley, R., Jones, D. L., Healey, J. R., Talboys, P. J., 2014. Feed the Crop Not the Soil: Rethinking Phosphorus Management in the Food Chain Environ. *Sci. Technol.* 48, 6523-6530.

Wu, Z., Dijkstra, P., Koch, G. W., Peñuelas, J., Hungate, B. A., 2011. Responses of terrestrial ecosystems to temperature and precipitation change: a meta-analysis of experimental manipulation. *Global Change Biology* 17, 927-942.

Xiang, D., Verbruggen, E., Hu, Y., Veresoglou, S. D., Rillig, M. C., Zhou, W., Xu, T., Li, H., Hao, Z., Chen, Y., Chen, B., 2014. Land use influences arbuscular mycorrhizal fungal communities in the farming–pastoral ecotone of northern China. *New Phytologist* 1-11, doi: 10.1111/nph.12961.

Zubek, S., Błaszkowski, J., Seidler-łozykowska, K., Baba, W., Mleczko, P., 2013. Arbuscular mycorrhizal fungi abundance, species richness and composition under the monocultures of five medicinal plants. *Acta Scientiarum Polonorum, Hortorum Cultus* 12, 5, 127-141.

EDITORS' CONTACT INFORMATION

Dr. Tapan Kumar Nath
Associate Professor
School of Biosciences,
University of Nottingham Malaysia Campus,
Selangor, Malaysia
Email: Tapan.Nath@nottingham.edu.my

Dr. Patrick O'Reilly
Socio-Economic Research Theme Coordinator
Crops for Future Research Centre
Selangor, Malaysia
Email: Patrick.O'Reilly@cffresearch.org

INDEX

#

A

B

C

D

G

H

I

J

N

O

P

Q

R

S

T

U

V

W

Y